Mathematical Problems and Proofs

Combinatorics, Number Theory, and Geometry

Mathematical Problems and Proofs

Combinatorics, Number Theory, and Geometry

Branislav Kisačanin
Delphi Delco Electronics Systems
Kokomo, Indiana

Plenum Press • New York and London

Library of Congress Cataloging-in-Publication Data

Kisačanin, Branislav.
Mathematical problems and proofs : combinatorics, number theory, and geometry / Branislav Kisačanin.
p. cm.
Includes bibliographical references and index.
ISBN 0-306-45967-1
1. Combinatorial analysis. 2. Set theory. 3. Number theory. 4. Geometry. I. Title.
QA164.K57 1998
511'.6--dc21 98-37206
CIP

This text was written using $\LaTeX 2_\varepsilon$,
while the figures were drawn in XFIG 2.1 or from MATLAB 4.2,
in the computer labs of the
Department of Electrical Engineering and Computer Science,
University of Illinois at Chicago

ISBN 0-306-45967-1

A Division of Plenum Publishing Corporation
233 Spring Street, New York, N.Y. 10013

http://www.plenum.com

10 9 8 7 6 5 4 3 2 1

Printed in the United States of America

To my love Saška
and
my hometown Novi Sad

Preface

For Whom Is this Book?

This book is written for those who enjoy seeing mathematical formulas and ideas, interesting problems, and elegant solutions.

More specifically it is written for talented high-school students who are hungry for more mathematics and undergraduates who would like to see illustrations of abstract mathematical concepts and to learn a bit about their historic origin.

It is written with that hope that many readers will learn how to read mathematical literature in general.

How Do We Read Mathematics Books?

Mathematics books are read with pencil and paper at hand. The reader sometimes wishes to check a derivation, complete some missing steps, or try a different solution.

It is often very useful to compare one book's explanation to another. It is also very useful to use the index and locate some other references to a theorem, formula, or a name.

Many people do not know that mathematics books are read in more than one way: The first reading is just browsing — the reader makes the first contact with the book. At that time the reader forms a first impression about contents, readability, and illustrations. At the second reading the reader identifies sections or chapters to read. After such second readings the reader may find the entire book interesting and worth reading from cover to cover. Every author aspires to be read in this way by more than just a few readers.

The reader should not expect to understand every proof or idea at once: It may be necessary to skip some details until other theorems or examples show the importance or further explain difficult parts. The reader will then discover that the previously unclear concepts are much easier to understand. Even when entire sections of a book are difficult to grasp, it is useful to skim them, so that at the next reading this material will be easier to understand.

What Does this Book Contain?

Besides many basic and some advanced theorems from combinatorics and number theory, this book contains more than 150 thoroughly solved examples and problems that illustrate theorems and ideas and develop the reader's problem-solving ability and sense for elegant solutions.

Historic notes and biographies of the four most important mathematicians of all time — Archimedes, Newton, Euler, and Gauss — will spark the reader's imagination and interest for mathematics and its history.

The main contents of the book are as follows:

Chapter 1 defines and explains set theory terminology and concepts. Several historically important examples are included.

Chapter 2 introduces the reader to elementary combinatorics. Before defining combinations and permutations, the reader is led through several examples to stimulate interest. Several illustrative examples in a separate section explain how to use the method of generating functions.

Chapter 3 introduces number theory, once the most theoretical of all mathematical disciplines and today the heart of cryptography. Among other topics the reader can find the Euclidean algorithm, Lamé's theorem, the Chinese remainder theorem, and a few words about the Fermat's last theorem.

Four appendixes at the end of the book provide additional information. Appendix A explains mathematical induction, an important mathematical tool. Appendix B provides many fascinating details and historic facts about four important mathematical constants: π, e, γ, and ϕ. Appendix C presents brief biographies of Archimedes, Newton, Euler, and Gauss, followed by a chronological list of many other important names from the history of mathematics. Appendix D gives the Greek alphabet.

Extensive references and index are provided for the benefit of the reader.

Acknowledgments

I would like to thank many people who supported me and helped me while I was writing this book. Above all my parents, Ljiljana and Miodrag; my brother, Miroslav; the love of my life, Saška; and my former and present professors, in particular Prof. Rade Doroslovački from the University of Novi Sad, and Prof. Gyan C. Agarwal from the University of Illinois at Chicago. Mrs. Apolonia Dugich and Dr. Miodrag Radulovački were great friends, and I wish to acknowledge their support, too. Finally I wish to thank Plenum and its mathematics editor, Mr. Thomas Cohn, for their interest in publishing my work, and their referees, whose names I will never know, for their comments and suggestions. Ms. Marilyn Buckingham did a wonderful job copyediting the manuscript. My wife, Saška, helped me compile the index.

Kokomo, IN Branislav Kisačanin

Contents

Key to Symbols

$\square$	End of proof or example
$a \in A$	a is in A
$b \notin A$	b is not in A
$A \subseteq B$	A is a subset of B
$A \subset B$	A is a proper subset of B
$A \cup B$	Set union
$A \cap B$	Set intersection
$A - B$	Set difference
$A \Delta B$	Symmetric set difference
$\lvert A \rvert$	Set cardinality
$A \times B$	Cartesian product of A and B
$\{1,2,3\}$	Set
$(1,2,3)$	Ordered triple
$\langle 1,2,3 \rangle$	Multiset
$\forall$	For all
$\exists$	There exists
$\exists!$	There exists only one
$p \wedge q$	Logical and
$p \vee q$	Logical or
$b \mid a$	b divides a
$b \nmid a$	b does not divide a
(a,b)	GCD of a and b
$[a,b]$	LCM of a and b
π	Ratio of the circumference to the diameter of a circle
e	Base of natural logarithms

ϕ	Golden section
γ	Euler's constant
f_n	Fibonacci number
F_n	Fermat number
M_p	Mersenne number
H_n	Harmonic number
$\varphi(n)$	Euler's function
$\mu(n)$	Möbius' function
$\pi(x)$	Number of primes $\leq x$
$\sigma(n)$	Sum of divisors of n
$\tau(n)$	Number of divisors of n
$\sum$	Sum
$\prod$	Product
lim	Limit
$=$	Equality
$\cong, \equiv$	Congruence in geometry
$\equiv$	Congruence in number theory
$\sim$	Asymptotic behavior
$\lfloor x \rfloor$	Largest integer $\leq x$
$n!$	n factorial
$\binom{n}{k}$	Binomial coefficient
$[AB]$	Segment
AB	Line
$\overrightarrow{AB}$	Vector
$\triangle ABC$	Triangle
$\angle ABC$	Angle

Mathematical Problems and Proofs

Combinatorics, Number Theory, and Geometry

1

Set Theory

Set theoretic terminology is used in all parts of mathematics, even in everyday language and life. In Chapter 1 we introduce the notation and terminology from set theory that are used in later chapters. We also show a few historically important examples that had a large impact on the development of mathematics.

1.1. Sets and Elementary Set Operations

Set and *set elements* are basic mathematical notions.

If a, b, and c are elements of the set A, then we write

$$A = \{a, b, c\}$$

In this case the element a is in A, while the element d is not in A. We write that as follows:

$$a \in A \qquad d \notin A$$

Instead of naming all elements of A by their names, it is often more convenient to define a set in the following analytic way:

$$A = \{x | P(x)\}$$

which means that A is the set of all elements having the property P.

EXAMPLE 1.1. The set of natural numbers less than 9 can be written in several equivalent ways, for example:

$$A = \{1,2,3,4,5,6,7,8\} \qquad A = \{1,2,\ldots,8\} \qquad A = \{n | n \in N \wedge n < 9\}$$

Some sets are used so often that there is a standard notation for them:

N Set of natural numbers
N_0 Set of natural numbers along with zero
Z Set of integers
Q Set of rational numbers
R Set of real numbers
C Set of complex numbers

NOTE: Authors often consider zero a natural number.

DEFINITION 1.1 (SUBSET). If for every element of A it is true that it is in B, too, we say that A is a subset of B, and write $A \subseteq B$.

DEFINITION 1.2 (EQUALITY OF SETS). Sets A and B are equal if $A \subseteq B$ and $B \subseteq A$. Then we write $A = B$.

Many proofs of equality of two sets proceed just as in this definition: First we prove that $A \subseteq B$, then that $B \subseteq A$.

DEFINITION 1.3 (PROPER SUBSET). If $A \subseteq B$ and $A \neq B$, we say that A is a proper subset of B, and write $A \subset B$.

EXAMPLE 1.2. If $A = \{1,2,3,4,5,6,7,8,9\}$ and $B = \{1,3,5,7,9\}$, then $B \subseteq A$. Since obviously $A \neq B$, we can also write $A \subset B$.

Through the following five definitions we introduce the most important set operations: union, intersection, difference, symmetric difference, and complement of a set. Each definition is illustrated by the corresponding Euler–Venn diagram in Figs. 1.1 and 1.2.

DEFINITION 1.4 (UNION). The union of sets A and B is the set of elements contained in at least one of these two sets:

$$A \cup B = \{x | x \in A \vee x \in B\}$$

EXAMPLE 1.3. If $A = \{1,2,3,4\}$ and $B = \{1,3,5,7,9\}$, then:

$$A \cup B = \{1,2,3,4,5,7,9\}$$

EXAMPLE 1.4. The union of the sets of odd and even integers is the set of all integers, i.e.,

$$Z_{\text{odd}} \cup Z_{\text{even}} = Z$$

EXAMPLE 1.5. The set of *rational* numbers Q consist of real numbers which can be represented as fractions of integers. Alternatively it is a set of reals with either a finite or periodic decimal representation. All other reals are called *irrational*; the set of irrational numbers is often denoted by I. Therefore we can write

$$Q \cup I = R$$

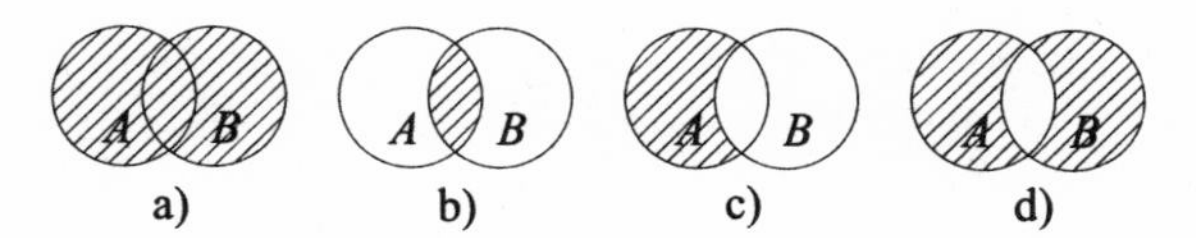

FIGURE 1.1. Euler–Venn diagrams of (a) union, (b) intersection, (c) difference, and (d) symmetric difference of sets.

DEFINITION 1.5 (INTERSECTION). The intersection of sets A and B is the set of elements contained in both of these sets:

$$A \cap B = \{x | x \in A \wedge x \in B\}$$

If the intersection of two sets is an empty set, i.e., if sets A and B do not have common elements, we say they are disjoint and write

$$A \cap B = \emptyset \qquad \text{or} \qquad A \cap B = \{\}$$

EXAMPLE 1.6. If $A = \{1,2,3,4\}$ and $B = \{1,3,5,7,9\}$, then:

$$A \cap B = \{1,3\}$$

EXAMPLE 1.7. The intersection of the sets of odd and even integers is the empty set, i.e., $Z_{\text{odd}} \cap Z_{\text{even}} = \emptyset$. Sets Q and I are also disjoint.

DEFINITION 1.6 (DIFFERENCE). The difference of sets A and B is the set of elements from A not contained in B:

$$A - B = \{x | x \in A \wedge x \notin B\}$$

EXAMPLE 1.8. If $A = \{1,2,3,4\}$ and $B = \{1,3,5,7,9\}$, then:

$$A - B = \{2,4\}$$

DEFINITION 1.7 (SYMMETRIC DIFFERENCE). The symmetric difference of sets A and B is the set of elements not contained in both A and B:

$$A \Delta B = (A \cup B) - (A \cap B)$$

EXAMPLE 1.9. If $A = \{1,2,3,4\}$ and $B = \{1,3,5,7,9\}$, then

$$A \Delta B = \{2,4,5,7,9\}$$

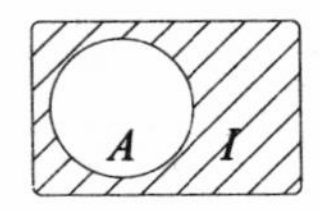

FIGURE 1.2. The complement of A with respect to I.

DEFINITION 1.8 (COMPLEMENT). If A is a subset of some set I, the complement of A with respect to I is the set of elements from I not contained in A:

$$\overline{A} = \{x | x \in I \wedge x \notin A\}$$

EXAMPLE 1.10. The complement of the set of even integers with respect to Z, the set of all integers, is the set of odd integers, i.e.,

$$\overline{Z_{\text{even}}} = Z_{\text{odd}}$$

EXAMPLE 1.11. Prove that $A - B = A \cap \overline{B}$.

SOLUTION: To prove this identity, the equality of these two sets, we must show that $x \in A - B$ if and only if $x \in A \cap \overline{B}$. Indeed:

$$x \in A - B \Leftrightarrow x \in A \wedge x \notin B \Leftrightarrow x \in A \wedge x \in \overline{B} \Leftrightarrow x \in A \cap \overline{B}$$

DEFINITION 1.9 (POWER SET). The set of all subsets of A is called the power set of A, and it is denoted by $P(A)$:

$$P(A) = \{X | X \subseteq A\}$$

Since by definition $\emptyset \subseteq A$ and $A \subseteq A$, then:

$$\emptyset \in P(A) \qquad A \in P(A)$$

NOTE: Later in the chapter on combinatorics, we prove that if A has n elements, then $P(A)$ has 2^n elements. For example, if $A = \{1,2,3\}$, then:

$$P(A) = \{\emptyset, \{1\}, \{2\}, \{3\}, \{1,2\}, \{1,3\}, \{2,3\}, \{1,2,3\}\}$$

DEFINITION 1.10 (SET PARTITION). Partition of a set A is a set of its nonempty, mutually disjoint subsets, whose union is A.

EXAMPLE 1.12. If $A = \{1,2,3\}$, then all partitions of A are $\{\{1\},\{2\},\{3\}\}$, $\{\{1\},\{2,3\}\}$, $\{\{2\},\{1,3\}\}$, $\{\{3\},\{1,2\}\}$, and $\{\{1,2,3\}\}$.

When defining a set, the order of its elements is irrelevant. It also does not matter if we list some element more than once, for example:

$$\{1,2,3\} = \{1,3,2\} = \{1,1,1,2,3\}$$

If we do care about the order and repetition of elements, we use ordered pairs, triples, etc., for example:

$$(1,2) \neq (2,1) \quad (1,1,2) \neq (1,2)$$

Here we define only the ordered pair because ordered triples, etc., are defined similarly.

DEFINITION 1.11 (ORDERED PAIR). The ordered pair (a,b) is defined as

$$(a,b) = \{\{a\},\{a,b\}\}$$

The element a is its first component, while b is its second component.

EXAMPLE 1.13. The ordered pair (a,b) is equal to another ordered pair (c,d) if and only if $a = c$ and $b = d$.

NOTE: In mathematics we often work with objects for which the order of their elements is irrelevant, but the repetition is not. To model such objects we use the so-called *multisets*. There will be more about them in Chapter 2.

1.2. Cartesian Product and Relations

DEFINITION 1.12 (CARTESIAN PRODUCT). The Cartesian product of sets A and B is the set of all ordered pairs in which the first component is from A and the second component is from B:

$$A \times B = \{(a,b) | a \in A \wedge b \in B\}$$

NOTE: The name Cartesian is derived from the Latin name of the French mathematician and philosopher René Descartes — Renatus Cartesius.

EXAMPLE 1.14. If $A = \{1,2,3\}$ and $B = \{7,9\}$, then:

$$A \times B = \{(1,7),(1,9),(2,7),(2,9),(3,7),(3,9)\}$$

DEFINITION 1.13 (RELATION). The relation ρ on set A is a subset of $A^2 = A \times A$:

$$\rho \subseteq A^2$$

If $(x,y) \in \rho$, we say that x is in relation with y. We also write $x \,\rho\, y$.

Relations can have many different properties. The following are the most important.

DEFINITION 1.14 (REFLEXIVITY). The relation ρ on A is reflexive if every element of A is in relation with itself:

$$\rho \text{ is reflexive} \Leftrightarrow (\forall x \in A) \quad x \,\rho\, x$$

DEFINITION 1.15 (SYMMETRY). The relation ρ on A is symmetric if for all $x,y \in A$, y is in relation with x whenever x is in relation with y:

$$\rho \text{ is symmetric} \Leftrightarrow (\forall x,y \in A) \quad x \,\rho\, y \Rightarrow y \,\rho\, x$$

DEFINITION 1.16 (ANTISYMMETRY). The relation ρ on A is antisymmetric if for all $x,y \in A$, $x \,\rho\, y$ and $y \,\rho\, x$ only when $x = y$:

$$\rho \text{ is antisymmetric} \Leftrightarrow (\forall x,y \in A) \quad (x \,\rho\, y \wedge y \,\rho\, x) \Rightarrow x = y$$

NOTE: There are relations that are neither symmetric nor antisymmetric. There are also relations that are both symmetric and antisymmetric. See Examples 1.27, 1.28, and 2.67.

DEFINITION 1.17 (TRANSITIVITY). The relation ρ on A is transitive if for all x,y, and $z \in A$, it follows from $x \,\rho\, y$ and $y \,\rho\, z$ that $x \,\rho\, z$:

$$\rho \text{ is transitive} \Leftrightarrow (\forall x,y,z \in A) \quad (x \,\rho\, y \wedge y \,\rho\, z) \Rightarrow x \,\rho\, z$$

The following three definitions introduce three important types of relations, which satisfy some of the preceding properties:

DEFINITION 1.18 (EQUIVALENCE RELATION). The relation ρ on A that is reflexive, symmetric, and transitive is called the equivalence relation.

EXAMPLE 1.15. If $A = \{1,2,3\}$ and $\rho = \{(1,1), (2,2), (3,3), (1,2), (2,1), (2,3), (1,3)\}$, then ρ is an equivalence relation on A. If A is given as $A = \{1,2,3,4\}$, then ρ is not reflexive, and therefore it is not an equivalence relation on A.

DEFINITION 1.19 (PARTIAL-ORDER RELATION). The relation ρ on A that is reflexive, antisymmetric, and transitive is called the partial-order relation.

DEFINITION 1.20 (SIMPLE-ORDER RELATION). The partial order relation ρ on A is a simple-order relation if every two elements of A are comparable, i.e., for every $x,y \in A$ we have either $x \,\rho\, y$, or $y \,\rho\, x$.

Simple-order relations are also called *linear-order relations*, while the corresponding sets are called *simply ordered sets* or *linearly ordered sets*.

EXAMPLE 1.16. If $A = \{1,2,3\}$ and $\rho = \{(1,1), (2,2), (3,3), (1,2), (2,3), (1,3)\}$, then ρ is a partial-order relation on A. This particular relation is usually denoted by $\leq$ (less then or equal). In this case every two elements of A are comparable, therefore $\leq$ is also a simple-order relation.

EXAMPLE 1.17. If $A = \{1,2,3,4,5,6\}$ and $\rho = \{(1,1), (2,2), (3,3), (4,4), (5,5), (6,6), (1,2), (1,3), (1,4), (1,5), (1,6), (2,4), (2,6), (3,6)\}$, then ρ is a partial-order relation on A. For example since 3 and 5 are not in relation (not comparable), this relation is not a simple-order relation. The reader may have recognized that this is the division relation.

1.3. Functions and Operations

A *function* or *mapping* is one of the most important concepts in mathematics.

DEFINITION 1.21 (FUNCTION). The function f of set X into set Y is a subset of the Cartesian product $X \times Y$ such that every $x \in X$ appears exactly once as the first component of the elements of f.

Symbolically we write

$$f : X \to Y \qquad x \mapsto f(x) \qquad \text{or} \qquad y = f(x)$$

Therefore a relation, i.e., a set of ordered pairs, is a function if and only if for no two ordered pairs their first components are equal and their second components differ.

EXAMPLE 1.18. If $A = \{1,2,3\}$ and $B = \{t,u,v,w\}$, then $f = \{(1,u), (2,w), (3,t)\}$ is a function, while $g = \{(1,u), (2,w)\}$ and $h = \{(1,u), (2,w), (3,t), (3,v)\}$ are not. Why? Because g does not specify the image for 3, while h specifies two of them instead of just one.

If $(x,y) \in f$, we say that x is an original, while y is its image. Set X is called the *domain*, while the set of all images $y = f(x)$, which is a subset of Y, is called the *range* or sometimes the *codomain*.

Functions f and g are equal if they have the same domain, and $f(x) = g(x)$ for every element x from the domain. Then we write $f = g$.

DEFINITION 1.22 (SURJECTION). The function f mapping X into Y such that the range is the whole set Y is called "onto" or a surjection.

DEFINITION 1.23 (INJECTION). The function f mapping X into Y such that no two originals have equal images is called "one-to-one" (1-1, for short) or an injection.

DEFINITION 1.24 (BIJECTION). The function f mapping X into Y that is both a surjection and an injection is called a "one-to-one correspondence" or a bijection.

Bijections have many important properties, and they are used very often. See the examples at the end of Chapter 1.

In the following we define binary operations and their properties.

DEFINITION 1.25 (BINARY OPERATION). The binary operation on the set A is a function of $A^2 = A \times A$ into A.

If a binary operation is denoted by $*$, then if $(a,b) \in A^2$ is mapped by $*$ on $c \in A$, we write

$$a * b = c$$

We now define the three basic properties of binary operations.

DEFINITION 1.26 (COMMUTATIVITY). The binary operation $*$ is commutative if:

$$(\forall a,b \in A) \quad a * b = b * a$$

DEFINITION 1.27 (ASSOCIATIVITY). The binary operation $*$ is associative if:

$$(\forall a,b,c \in A) \quad (a*b)*c = a*(b*c)$$

DEFINITION 1.28 (DISTRIBUTIVITY). The binary operation $*$ is distributive with respect to the binary operation $\circ$ if:

$$(\forall a,b,c \in A) \quad (a \circ b)*c = (a*c)\circ(b*c) \land c*(a \circ b) = (c*a)\circ(c*b)$$

1.4. Cardinality

DEFINITION 1.29 (INFINITE SET). A set is infinite if it can be bijectively mapped onto some of its proper subsets.

DEFINITION 1.30 (FINITE SET). A set is finite if it is not infinite.

EXAMPLE 1.19 (SET N IS INFINITE). The set of even natural numbers is a proper subset of the set of natural numbers N. Since $f(n) = 2n$ is a bijection, the set N is infinite. The fact that there is a one-to one correspondence between these two sets is an apparent paradox noted by Galileo in 1638.

EXAMPLE 1.20 (EUCLID'S THEOREM ON PRIMES). In the ninth book of his *Elements*, Euclid gives the following proof of the infiniteness of the set of primes.

Assume there are only finitely many primes, $p_1, p_2, \ldots, p_n$ and let p_n be the greatest among them. Consider

$$P = p_1 p_2 \ldots p_n + 1$$

which is obviously $P > p_n$. There are two possibilities for P:

- P is a prime. This contradicts the assumption that p_n is the greatest prime.
- P is a composite. This contradicts the assumption that $p_1, p_2, \ldots, p_n$ are all primes. Dividing P by any of these yields remainder 1; i.e., P has prime factors that differ from $p_1, p_2, \ldots, p_n$.

This proves the infiniteness of the set of primes.

NOTE: Proofs by *contradiction* are very common in mathematics: We first assume that a statement is true, then we show that this assumption leads to a contradiction.

DEFINITION 1.31 (CARDINALITY). Sets A and B have the same cardinalities and we write $|A| = |B|$ if there exists a bijection $f : A \to B$.

The cardinality of A equals n, i.e., $|A| = n$, if and only if there exists a bijection $f : A \to \{1,2,\ldots,n\}$.

The cardinalities of different infinite sets are not all equal. Hence the cardinality of N is denoted by $\aleph_0$ (read: *aleph-zero*; $\aleph$ is the first letter of the Hebrew alphabet), while the cardinality of R is denoted by c (from the Latin *continuum*). In what follows, we see where differences in cardinal numbers come from, and how they are manifested. These numbers are often called *transfinite*.

DEFINITION 1.32 (COUNTABLE SET). An infinite set A is countable if there is a bijection $f : A \to N$.

Equivalently an infinite set A is countable if its elements can be arranged in a sequence $a_1, a_2, a_3, \ldots$

DEFINITION 1.33 (UNCOUNTABLE SET). An infinite set is uncountable if it is not countable.

EXAMPLE 1.21 (SET Z IS COUNTABLE). The set of all integers Z is countable because integers can be arranged in a sequence:

$$0, 1, -1, 2, -2, 3, -3, 4, -4, \ldots$$

EXAMPLE 1.22 (SET Q IS COUNTABLE). The proof by Cantor, one of the founders of modern set theory, that the set of rational numbers can be written as a sequence, i.e., Q is countable, follows.

Every rational number can be represented as a fraction of two relatively prime integers p/q. First write all rational numbers where $p+q = 1$, then add those where $p+q = 2$ if they are not already in the sequence, then those where $p+q = 3$ only if these are not already included in the sequence, etc.,

$$\frac{0}{1}, \frac{1}{1}, \frac{1}{2}, \frac{2}{1}, \frac{1}{3}, \frac{3}{1}, \frac{1}{4}, \frac{2}{3}, \frac{3}{2}, \frac{4}{1}, \ldots$$

EXAMPLE 1.23 (SET R IS UNCOUNTABLE). This proof was also given by Cantor in 1874, and it is called Cantor's diagonal procedure in his honor. Without loss of generality, we show that the set $A = \{x | x \in R \wedge 0 < x < 1\}$ is uncountable, where A is an example of an *open interval*, written as $A = (0, 1)$.

Every real number from $(0,1)$ can be uniquely written as a decimal number with infinitely many digits different from zero. Exception are not even the numbers with a finite representation, because they can be written with infinitely many nines, for example:

$$0.123 = 0.12299999\ldots$$

Assume the interval $(0,1)$ to be countable. Then all numbers $0 < x < 1$ are in a sequence

$$\begin{array}{l} 0.a_{11}a_{12}a_{13}a_{14}\ldots \\ 0.a_{21}a_{22}a_{23}a_{24}\ldots \\ 0.a_{31}a_{32}a_{33}a_{34}\ldots \\ 0.a_{41}a_{42}a_{43}a_{44}\ldots \\ 0.a_{51}a_{52}a_{53}a_{54}\ldots \\ \vdots \end{array}$$

But the number $x = 0.x_1x_2x_3\ldots$ defined by $0 \neq x_k \neq a_{kk}$ $(k = 1,2,3,\ldots)$ is not in that sequence. For any $k \in N$, x is not the kth number in the sequence because by definition, the kth digits of x and $0.a_{k1}a_{k2}a_{k3}\ldots$ are different. □

Since R is uncountable, i.e., no bijection between N and R exists, we can write $\aleph_0 \neq c$. In fact since $N \subset R$, we can write $\aleph_0 < c$.

For a long time mathematicians did not know if there were a set with cardinality between $\aleph_0$ and c. The answer to that so-called *continuum hypothesis* was given in 1939 by Kurt Gödel who showed that the continuum hypothesis does not contradict the axioms of set theory and in 1964 by Paul Cohen, who showed that it also does not follow from them. In other words the existence of the set A such that $\aleph_0 < |A| < c$ can be taken as a new and independent axiom of set theory.

The continuum hypothesis is an example of Gödel's famous incompleteness theorem from 1931 which states that in every consistent mathematical system, there are theorems which are neither provable nor disprovable. This is similar to a paradox discovered in Ancient Greece and is usually attributed to Epimenides of Crete (sixth century B.C.) or to Eubulides of Miletus (fourth century B.C.): "What I am now saying is a lie." If this statement is true, it must be false, and vice versa, if it is false, it must be true. Therefore it is neither false nor true.

1.5. Problems

EXAMPLE 1.24 (PROPERTIES OF SET OPERATIONS). The following properties of set operations are easy to prove. Let A, B, and C be arbitrary sets, then:

$A \cup A = A$	Idempotency of union
$A \cap A = A$	Idempotency of intersection
$A \cup B = B \cup A$	Commutativity of union
$A \cap B = B \cap A$	Commutativity of intersection
$(A \cup B) \cup C = A \cup (B \cup C)$	Associativity of union
$(A \cap B) \cap C = A \cap (B \cap C)$	Associativity of intersection
$(A \cap B) \cup C = (A \cup C) \cap (B \cup C)$	Distributivity of $\cup$ with respect to $\cap$
$(A \cup B) \cap C = (A \cap C) \cup (B \cap C)$	Distributivity of $\cap$ with respect to $\cup$
$A \subset B \Rightarrow A \cup B = B$	
$A \subset B \Rightarrow A \cap B = A$	
$A \cup (A \cap B) = A$	
$A \cap (A \cup B) = A$	
$\overline{\overline{A}} = A$	Involutivity of complement
$\overline{A \cup B} = \overline{A} \cap \overline{B}$	De Morgan's law
$\overline{A \cap B} = \overline{A} \cup \overline{B}$	De Morgan's law

EXAMPLE 1.25. Among 50 participants in the Mathematical Olympiad, 33 like chicken, 20 like pork, while 18 like beef. If no competitor likes all three kinds of meat, eight competitors like both chicken and pork, nine like pork and beef, and seven like chicken and beef, find how many of them are vegetarians.

SOLUTION: Problems like this are usually solved using Euler–Venn diagrams. It is easy to see from Fig. 1.3 that there are three vegetarians among the competitors.

EXAMPLE 1.26. Show that the Cartesian product is not a commutative operation.

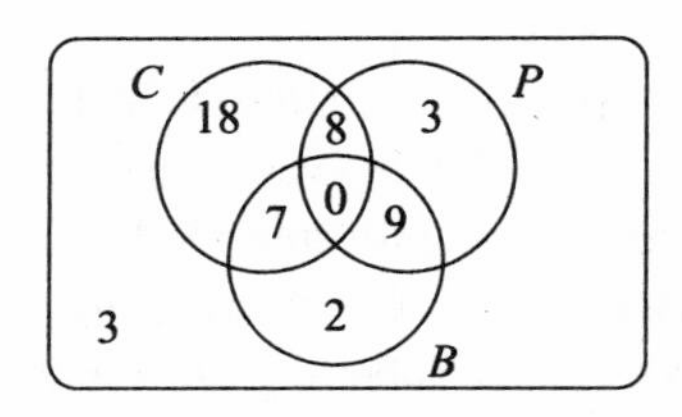

FIGURE 1.3. Euler–Venn diagram for finding the number of vegetarians.

SOLUTION: It is enough to find a pair of sets A and B such that $A \times B \neq B \times A$:

$$A = \{1,2\},\ B = \{3\} \ \Rightarrow\ A \times B = \{(1,3),(2,3)\} \neq \{(3,1),(3,2)\} = B \times A$$

NOTE: This is a typical solution by finding a *counterexample*.

EXAMPLE 1.27. Show that the relation $\rho = \{(1,1),\ (2,2)\}$ on $A = \{1,2,3\}$ is not reflexive but symmetric, antisymmetric, and transitive.

EXAMPLE 1.28. Show that $\rho = \{(1,2),\ (1,3),\ (2,1),\ (2,3)\}$ defined on $A = \{1,2,3\}$ has none of the properties in Example 1.27.

EXAMPLE 1.29 (EQUIVALENCE CLASSES). Consider an equivalence relation on A denoted by $\sim$ (read: *tilde*). Denote by $[x]$ a subset of A that contains all elements from A in relation with $x \in A$ and these elements only, i.e.,

$$[x] = \{y | y \in A \wedge x \sim y\}$$

The set $[x]$ is called the *equivalence class* of x. Since $\sim$ is an equivalence relation, i.e., it is reflexive, symmetric, and transitive, we easily see that:

$$[x] = [y] \ \Leftrightarrow\ x \sim y$$

This implies that every two equivalence classes are either disjoint or equal to each other. The union of all equivalence classes is obviously A. The set of all equivalence classes is called the *quotient set*, and it is denoted by $A/\sim$. Since the equivalence classes are disjoint and their union is A, each equivalence relation describes one partition of A, and vice versa, every partition of A defines one equivalence relation.

EXAMPLE 1.30 (FOR EXAMPLE 1.29). Define the relation "has the same remainder when divided by 5 as" on the set of integers Z. There are five equivalence classes (in this particular case called *residue classes*): [0], [1], [2], [3], and [4]:

$$[0] = \{\ldots,-5,0,5,\ldots\} \quad [1] = \{\ldots,-4,1,6,\ldots\} \quad [2] = \{\ldots,-3,2,7,\ldots\}$$
$$[3] = \{\ldots,-2,3,8,\ldots\} \quad [4] = \{\ldots,-1,4,9,\ldots\}$$

It is obvious that $[0] \cup [1] \cup [2] \cup [3] \cup [4] = Z$ and $[i] \cap [j] = \emptyset \ (i \neq j)$. This relation is usually denoted by $\equiv$, and we write, e.g.,

$$16 \equiv 1 \pmod 5 \qquad \text{(read: 16 } \textit{is congruent to } 1 \textit{ modulo } 5\text{)}$$

The quotient set now is $(Z/\equiv) = \{[0],[1],[2],[3],[4]\}$.

EXAMPLE 1.31 (IRRATIONAL NUMBERS). Until the Pythagoreans, students and followers of Pythagoras, discovered that the diagonal of a square is not commensurable to the side of the square or in other words that $\sqrt{2}$ is not a rational number, ancient mathematicians were content with rational numbers, i.e., numbers that can be written as integer fractions.

We prove here that $\sqrt{2}$ is irrational as the Pythagoreans did by contradiction. Suppose $\sqrt{2}$ is rational; i.e., it can be written as a fraction of integers:

$$\sqrt{2} = \frac{a}{b}$$

Assume also that a and b are such that they do not have common factors. Assuming all this, and squaring the previous equality, we obtain

$$a^2 = 2b^2$$

This implies a is an even number, i.e., $a = 2a_1$. But:

$$b^2 = 2a_1^2$$

This implies b is an even number, too, which contradicts our assumption that a and b have no common factors.

Therefore we find that it is impossible to write $\sqrt{2}$ as a fraction of integers; i.e., $\sqrt{2}$ is irrational.

EXAMPLE 1.32 (TRANSCENDENTAL NUMBERS). Among irrationals, too, there are different kinds of numbers. Irrationals that can be defined as the roots* of polynomials with integer coefficients are called *algebraic numbers*. The golden section ϕ is one of the solutions of $x^2 - x - 1 = 0$, hence ϕ is an algebraic number. Irrational numbers that are not algebraic are called *transcendental numbers*.

In 1873 Hermite proved that e is a transcendental number, and in 1882 Lindemann showed the same for π, thus also proving that, using only a ruler and compass, it is impossible to construct the square whose area is the same as the given circle. This is the so-called problem of *squaring a circle*, which remained unsolved since antiquity. It is still not known whether some important

*The following are synonyms: the roots of the polynomial $P(x)$, the zeros of the polynomial $P(x)$, and the solutions of the equation $P(x) = 0$.

mathematical constants are rational or irrational, let alone whether they are algebraic or transcendental (if they are irrational). One such numbers is Euler's constant

$$\gamma = \lim_{n\to\infty}\left(1+\frac{1}{2}+\frac{1}{3}+\ldots+\frac{1}{n}-\ln n\right)$$

See Appendix B for more about these important numbers.

EXAMPLE 1.33 (RATIONAL OR NOT?). If a and b are irrational, can a^b be a rational number?

SOLUTION: Yes! Consider $\sqrt{2}^{\sqrt{2}}$. If it is rational, then an effective example is $a = b = \sqrt{2}$. If it is irrational, then take $a = \sqrt{2}^{\sqrt{2}}$ and $b = \sqrt{2}$. Then:

$$a^b = \left(\sqrt{2}^{\sqrt{2}}\right)^{\sqrt{2}} = (\sqrt{2})^2 = 2$$

EXAMPLE 1.34 (ABORIGINAL ELECTIONS). The Aborigines of Australia pick for their head the man with the largest flock of sheep. But since in their language and culture there are no numbers larger than 20, they have an ingenious election system: One sheep from each flock of the two finalists is taken through a gate, until it is determined which man has the larger flock: Mapping at work!

EXAMPLE 1.35. Let A be a set with n elements; i.e., let $|A| = n$. Show that the number of subsets of A having k $(0 \le k \le n)$ elements is equal to the number of subsets of A having $(n-k)$ elements.

SOLUTION: To an arbitrary k-element set $B \subset A$ we can uniquely ascribe the $(n-k)$-element set $A - B \subset A$. This mapping is a bijection from the set of all k-element subsets of A, $P_k(A)$, onto the set of all $(n-k)$-element subsets of A, $P_{n-k}(A)$. Hence we find

$$|P_k(A)| = |P_{n-k}(A)|$$

Also see Example 2.27.

EXAMPLE 1.36. Let $a \in A$. Are there more subsets of A containing a or those not containing it?

SOLUTION: The function f that maps every subset B not containing a onto the subset $B \cup \{a\}$ is obviously a bijection; therefore the cardinalities of sets of these subsets are equal. This solution does not depend on whether or not A is finite.

EXAMPLE 1.37. Let $X \subset A$. Are there more subsets of A that contain X are disjoint with it?

RESULT: The cardinalities are equal again; the whole problem is very similar to the one in Example 1.36.

EXAMPLE 1.38. Let X be a subset of a finite set A and let $|X| > 1$. Are there more subsets of A which contain X, or those which do not?

RESULT: Since not all subsets of A that do not contain X are disjoint with it, more subsets do not contain X than do.

EXAMPLE 1.39. Show that for an arbitrary set A, the number of subsets with an even number of elements equals the number of subsets with an odd number of elements.

SOLUTION: Let us pick one element from A and denote it by a. Define a function f from the set of even-numbered subsets onto the set of odd-numbered subsets as follows: If an even-numbered subset B contains a, let $f(B) = B - \{a\}$, and if $a \notin B$, let $f(B) = B \cup \{a\}$. It is easy to see that f is a bijection. This completes the proof.

Also see Example 2.54.

2

Combinatorics

Chapter 2 discusses the basic notions of combinatorics. At the beginning, our main task is to solve enumeration problems without considering permutations, combinations, etc. That is, we try to find the best way of enumerating objects and their arrangements. Once we have a reasonable amount of experience with such problems, we define and use combinatorial terminology. Even then we sometimes find it easier to solve problems simply by counting.

2.1. Four Enumeration Principles

To answer such questions as *How many ways are there to give* 30 *books to seven friends?* as well as much more difficult questions, we use the following enumeration rules and principles:

Let A and B be sets with m and n elements, respectively; i.e., let

$$|A| = m \qquad |B| = n$$

Rule of product. *The number of ways of forming an ordered pair* (a,b) *such that* $a \in A$ *and* $b \in B$*; equals* $m \cdot n$*. In other words:*

$$|A \times B| = m \cdot n$$

Rule of sum. *If sets A and B are disjoint, then the number of ways of picking one element from their union equals the sum* $m+n$*. In other words:*

$$A \cap B = \emptyset \Rightarrow |A \cup B| = m+n$$

Principle of inclusion–exclusion. *In general when A and B are not necessarily disjoint, the following is true:*

$$(A \cap B = C \quad |C| = p \geq 0) \Rightarrow |A \cup B| = m+n-p$$

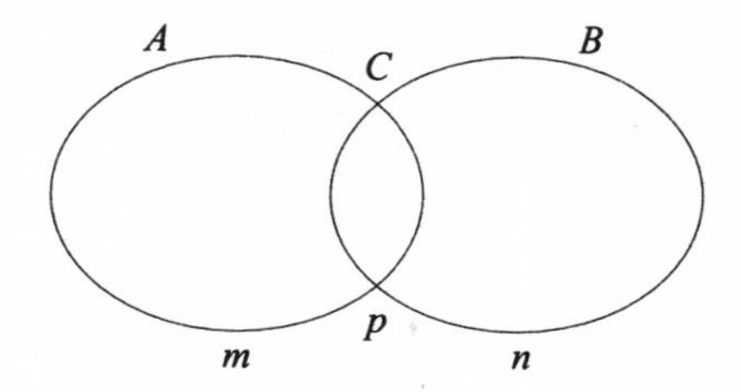

FIGURE 2.1. With the principle of inclusion–exclusion. If $|A| = m$, $|B| = n$, and $|A \cap B| = p$, then $|A \cup B| = m + n - p$.

It is easy to see that the principle of inclusion–exclusion is a consequence of the rule of sum. Consider Fig. 2.1. First of all note that sets $B - C$ and C are disjoint and their union equals B. Hence $|B| = |B - C| + |C|$, i.e., $|B - C| = |B| - |C| = n - p$. Since sets A and $B - C$ are disjoint and their union equals the union of A and B, we find that:

$$\begin{aligned} |A \cup B| &= |A \cup (B - C)| \\ &= m + n - p \end{aligned}$$

In other words the union of two sets in the general case does not have $m + n$ elements because this sum counts the common elements of A and B twice. To obtain the correct result, we must subtract p, the number of elements in $A \cap B$.

Dirichlet's principle. *If n disjoint sets contain $n + 1$ elements, at least one of them has more than one element.*

Although these principles are very simple, they are used to solve even very difficult problems.

2.2. Introductory Problems

Many books begin with simple combinatorial problems grouped according to what formula is used to solve them. Our aim here is to try to develop the reader's ability to enumerate objects and their arrangements, not to recognize which formula to apply. Therefore in Section 2.2, we show several problems and their solutions without mentioning permutations and combinations even once.

EXAMPLE 2.1. How many elements are there in the set given by $A = \{n, n+1, \ldots, n^2\}$?

SOLUTION: It is obvious that $|A| = n^2 - n + 1$, but let us try to justify this answer. What does it actually mean to *enumerate* the elements of a set? If A has $r \in N$ elements, then each element can be assigned an order, a number from the set $N_r = \{1, 2, 3, \ldots, r\} \subset N$.

In our problem, we assign order 1 to n, order 2 to $n+1$, and so on. The question now is what is the order of n^2? We notice that in this case, the difference between the element of A and its order is always $n-1$. Therefore the order of n^2 is $n^2 - (n-1) = n^2 - n + 1$, which implies that $|A| = n^2 - n + 1$.

EXAMPLE 2.2. A certain island is home to 510 seals. Suppose each seal has 10 or more mustaches, but not more than 30. Prove that among these 510 seals at least 25 of them have an equal number of mustaches.

SOLUTION: If we divide the seals into 21 groups according to how many mustaches they have (note that 21 is the cardinality of the set $\{10, 11, \ldots, 30\}$) and assume that each group has ≤ 24 seals, then there are $\leq 21 \cdot 24 = 504 < 510$ seals on the island. Hence, at least one group has 25 or more seals.

EXAMPLE 2.3. How many seven-digit phone numbers begin with 432 and end with 3 or 5?

SOLUTION: The fourth digit can be selected from 10 possible choices: $0, 1, 2, \ldots, 9$. The same is true for the fifth and the sixth digit. The seventh digit can be picked in two ways: it can be either 3 or 5. Therefore, according to the rule of product, the number of such phone numbers is $10 \cdot 10 \cdot 10 \cdot 2 = 2000$.

EXAMPLE 2.4. There are n points given in a plane, and no three of these are collinear, i.e., no three lie on the same line. How many lines are defined by these n points?

SOLUTION: Since each pair of different points defines a line, in this problem we are actually interested in counting all pairs of different points that can be formed from the n given points. The first point can be picked in n different ways, while the other can be picked in $(n-1)$ ways since it must differ from the first. Hence the number of pairs is $n(n-1)$.

Please note that every line formed in this way appears twice because, for example, ordered pairs (A,B) and (B,A) define the same line: $AB \equiv BA$. Therefore the number of lines is twice as small as the number of ordered pairs; it equals $n(n-1)/2$.

EXAMPLE 2.5. Similarly if n points are given in space and among them no four coplanar, i.e., no four lie in the same plane, there are $n(n-1)(n-2)/6$ planes defined by these n points.

NOTE: In Examples 2.4 and 2.5 we reduced the problems to counting the two- or three-element subsets of the n-element sets. Since the order of the elements of a (sub)set is irrelevant, we can uniquely assign $\{A, B\}$ to the line $AB \equiv BA$ and $\{A, B, C\}$ to the plane $\pi_{ABC} \equiv \pi_{ACB} \equiv \ldots \equiv \pi_{CBA}$.

EXAMPLE 2.6. How many seven-digit phone numbers begin with 215-2 if the last three digits must differ among themselves; they cannot be 0, 2, or 5; and the last digit cannot be 1?

SOLUTION 1: Let us try to solve this problem as follows: We cannot change the first four digits; these are fixed. The fifth digit can be selected from the set $\{1,3,4,6,7,8,9\}$, i.e., in seven different ways. The sixth digit must differ from the fifth, so it can be picked in six ways. The last digit must be different from the fifth and the sixth, so without the last condition, we could pick it in five ways. Due to the last condition we cannot select the last digit in four ways, because if the fifth or sixth digit were 1, the last digit can be selected in five ways, not four.

Let us divide the set of all phone numbers satisfying these conditions into three disjoint sets: the set of numbers having neither a fifth nor sixth digit equal to 1, the set of numbers having 1 at the fifth place, and the set of numbers having 1 at the sixth place.

In the first set, there are $6 \cdot 5 \cdot 4$ choices. In the second set the first five digits are fixed, so there are $6 \cdot 5$ choices. Similarly in the third set, there are $6 \cdot 5$ choices. Therefore $6 \cdot 5 \cdot 4 + 1 \cdot 6 \cdot 5 + 6 \cdot 1 \cdot 5 = 180$ phone numbers have the required properties.

SOLUTION 2: The solution can be obtained much more easily if we begin with the last digit, which can be picked from $\{3,4,6,7,8,9\}$, i.e., in 6 ways. Once it is selected, we can select the sixth digit in six ways, too, because although we cannot repeat the last digit, we can use digit 1. The fifth digit can be selected in five ways. The result is the same as before: $6 \cdot 6 \cdot 5 = 180$.

SOLUTION 3: Let us consider the third way to solve this problem. From the total number of phone numbers that can be formed using different digits from $\{1,3,4,6,7,8,9\}$ at their last three places, we subtract the number of phone numbers having 1 at the last place: $7 \cdot 6 \cdot 5 - 6 \cdot 5 = 180$.

EXAMPLE 2.7. From the set containing n arbitrary natural numbers $\{a_1,\ldots,a_n\}$ we can select a subset in which the sum of all elements is divisible by n. Prove this.

SOLUTION: Let us consider n subsets:

$$\{a_1\},\{a_1,a_2\},\ldots,\{a_1,a_2,\ldots,a_n\}$$

First calculate the sums in each of these subsets and in the remainders after division by n. If some of these remainders is 0, we have the subset we seek. If none of these is 0 then, according to Dirichlet's principle, among these n subsets there are two with equal remainders. [If none of these has remainder 0, then n remainders are to be distributed in $(n-1)$ residue classes.] Let these two subsets be $\{a_1,a_2,\ldots,a_r\}$ and $\{a_1,a_2,\ldots,a_s\}$, where, e.g., $r<s$. Then:

$$a_1+a_2+\ldots+a_s-(a_1+a_2+\ldots+a_r)=a_{r+1}+a_{r+2}+\ldots+a_s$$

is divisible by n, and $\{a_{r+1},a_{r+2},\ldots,a_s\}$ is the subset we seek.

EXAMPLE 2.8 (NUMBER OF DIVISORS). How many different divisors, including itself and 1, does 2520 have?

SOLUTION: Canonical decomposition of 2520 is $2520=2^3\cdot 3^2\cdot 5\cdot 7$, so all divisors of 2520 can be written as $2^a\cdot 3^b\cdot 5^c\cdot 7^d$, where $a,b,c,d\geq 0$; each of these is less than or equal to the corresponding exponent in the decomposition of 2520. In other words, $0\leq a\leq 3$, $0\leq b\leq 2$, $0\leq c\leq 1$, $0\leq d\leq 1$. Hence a can be picked in four ways, b in three, c in two, and d in two. Therefore, the number of divisors of 2520 is $4\cdot 3\cdot 2\cdot 2=48$.

NOTE: In number theory the number of divisors is denoted by $\tau(n)$ or sometimes by $d(n)$. If $p_1,p_2,\ldots,p_r$ are the prime factors of n, then:

$$n=p_1^{\alpha_1}p_2^{\alpha_2}\ldots p_r^{\alpha_r}\ \Rightarrow\ \tau(n)=(\alpha_1+1)(\alpha_2+1)\ldots(\alpha_r+1)$$

EXAMPLE 2.9. A code for a safe is a five-digit number that can have a 0 in the first place as well as in any other place. How many codes are there whose digits form an increasing sequence?

SOLUTION: Consider codes composed of different digits without the increasing order restriction. Suppose we select five out of ten possible digits to form these codes. Among all $5 \cdot 4 \cdot 3 \cdot 2 \cdot 1 = 120$ codes that can be formed using these five digits, only one code has its digits in increasing order.

This observation is already an important step toward the solution. Let M be the solution, the number of length-5 increasing sequences.

The total number of codes with different digits is $10 \cdot 9 \cdot 8 \cdot 7 \cdot 6 = 30240$, but also $120M$, hence:

$$M = \frac{10 \cdot 9 \cdot 8 \cdot 7 \cdot 6}{5 \cdot 4 \cdot 3 \cdot 2 \cdot 1} = \frac{30240}{120} = 252$$

EXAMPLE 2.10. A code for a safe is a five-digit number that can have a 0 at the first place. How many codes have *exactly* one digit 7?

SOLUTION: There are $5 \cdot 9^4 = 52488$ codes with exactly one digit 7. Actually, there are $9 \cdot 9 \cdot 9 \cdot 9 = 9^4$ codes with a digit 7 in the first place. The same is true for codes with 7 in the second, third, fourth, and fifth places. The total number is therefore $5 \cdot 9^4$. We obtain the same result if we say there are five choices for placing 7 and nine choices for each of the remaining four places.

EXAMPLE 2.11. A code for a safe is a five-digit number that can have a 0 at the first place. How many codes have *at least* one digit 7?

SOLUTION: There are $5 \cdot 9^4$ with exactly one 7. There are $10 \cdot 9^3$ codes with exactly two digits 7 because two places containing 7 can be selected in ten ways. Similarly there are $10 \cdot 9^2$ codes with exactly three 7s, $5 \cdot 9$ codes with exactly four 7s, and only one code with all five digits equal to 7. The total is

$$5 \cdot 9^4 + 10 \cdot 9^3 + 10 \cdot 9^2 + 5 \cdot 9 + 1 = 40951$$

The more elegant way of solving this problem involves subtracting the number of codes not having any digits equal to 7 from the total number of codes:

$$10^5 - 9^5 = 40951$$

NOTE: The following equality, obtained by comparison of the two solutions,

$$10^5 = 9^5 + 5 \cdot 9^4 + 10 \cdot 9^3 + 10 \cdot 9^2 + 5 \cdot 9 + 1$$

is a special case of Newton's binomial expansion.

EXAMPLE 2.12. A can of red paint is spilled over a white plane. Show that this red and white plane contains two points of the same color whose distance is exactly 1 cm.

SOLUTION: Consider the vertices of an equilateral triangle whose side is 1 cm. Each of these three points is either red or white; hence two of them have the same color.

EXAMPLE 2.13. Let us consider a chess board with a knight moving on it. Can the knight start from the lower left corner (A-1), visit every field on the board exactly once, and end in the upper right corner (H-8)?

SOLUTION: The answer is no, and this is why. The first field (A-1) is black, the second, according to rules governing the knight's motion, must be white, the third is then black, and so on. We thus observe that even fields must be white, including the last, sixty-fourth field. But H-8 is black, so it cannot be the last field. □

From previous examples we see that parts, sometimes even an entire problem, can be reduced to counting the number of subsets or ordered k-tuples formed by elements of some set. For that reason mathematicians introduced terms like combinations and permutations. Before defining these, let us consider a few examples using only set-theoretic terminology.

EXAMPLE 2.14. How many subsets does $A = \{a_1, a_2, \ldots, a_n\}$ have?

SOLUTION: Consider an arbitrary subset of A. Each of n elements of A is either in that subset or not. Thus according to the rule of product, the number of different subsets of A is

$$\underbrace{2 \cdot 2 \cdot \ldots \cdot 2}_{n} = 2^n$$

EXAMPLE 2.15. How many ordered k-tuples can be formed from different elements of $A = \{a_1, a_2, \ldots, a_n\}$?

SOLUTION: The first component of the ordered k-tuple can be filled by one of n elements of A, the second by any of the remaining $n-1$ elements, etc. The kth component can be picked from the last remaining $n-k+1$ elements of A. Hence the total number of ordered k-tuples formed from different elements of A is

$$n(n-1)(n-2)\ldots(n-k+1) = \frac{n!}{(n-k)!}$$

where $r! = r(r-1)\dots 1$ is the factorial of r (read: *r factorial*). For this and similar formulas to be valid when $n = k$, the convention is that $0! = 1$.

EXAMPLE 2.16. How many ordered k-tuples can be formed from the elements of the set $A = \{a_1, a_2, \dots, a_n\}$ if repetition is allowed?

SOLUTION: Since repetition is allowed, each component can be selected in any of n ways; hence, the number of such ordered k-tuples is

$$\underbrace{n \cdot n \cdot \ldots \cdot n}_{k} = n^k$$

EXAMPLE 2.17. How many k-element subsets does the n-element set A have?

SOLUTION: Assume without any loss of generality that the set A is $A = \{1, 2, \dots, n\}$. As in Example 2.9, there are as many k-element subsets of A as there are increasing sequences made from its elements. Let the final answer be M. From each of M increasing length-k sequences, we can form $k(k-1)\dots 1 = k!$ ordered k-tuples. Hence the number of all ordered k-tuples whose components are different elements of A equals $k! \cdot M$. On the other hand, as in Example 2.15, their number equals

$$\frac{n!}{(n-k)!}$$

Therefore:

$$M = \frac{n!}{k!\,(n-k)!}$$

We see later that instead of this clumsy quotient of factorials, we write

$$M = \binom{n}{k}$$

which reads *n choose k*.

It is not a coincidence that the same symbol is used for the binomial coefficients, and we learn why very soon. □

Unlike with ordered k-tuples, the order of elements for sets is irrelevant, for example:

$$(1,2,3) \neq (1,3,2) \qquad \{1,2,3\} = \{1,3,2\}$$

Besides $\{1,2,2\} = \{1,2\}$.

However in some cases we have objects whose order is irrelevant, but their repetition may occur. In such cases we must be careful with the notation we use or even better to introduce the *multisets*.

Unlike with ordered k-tuples, order is irrelevant for multisets. At the same time, unlike with sets, repetition is allowed. We use $\langle a_1, a_2, \ldots, a_n \rangle$ to denote a multiset, for example, $\langle 1,2,2,3 \rangle \neq \langle 1,2,3 \rangle$ and $\langle 1,2,2,3 \rangle = \langle 1,2,3,2 \rangle$. Hence instead of asking the wrong question, How many k-element subsets can be formed from the elements of an n-element set A if repetition is allowed? we ask the following question in Example 2.18.

EXAMPLE 2.18. How many k-element multisets can be formed from the elements of $A = \{a_1, a_2, \ldots, a_n\}$?

SOLUTION: Every k-element multiset consisting of elements from A can be uniquely represented as a sequence of k zeros and $n-1$ vertical lines. The number of zeros to the left from the first line represents the number of repetitions of a_1, the number of zeros between the first and second lines represents the number of repetitions of $a_2, \ldots$; the number of zeros to the right from the last vertical line represents the number of repetitions of a_n.

For example if $n = 7$ and $k = 4$, the multiset $\langle a_1, a_1, a_3, a_6 \rangle$ can be represented as $00||0|||0|$.

If the zeros differ among themselves and the lines among themselves, the number of different sequences of zeros and lines would be $(n-1+k)!$ But since the zeros are indistinguishable (as well as the lines) the number of different sequences of zeros and lines is

$$\frac{(n-1+k)!}{(n-1)!k!} = \binom{n+k-1}{k}$$

EXAMPLE 2.19. Let us consider how many different ordered n-tuples can be formed from the elements of the multiset:

$$A = \langle \underbrace{a_1, \ldots, a_1}_{n_1}, \underbrace{a_2, \ldots, a_2}_{n_2}, \ldots, \underbrace{a_r, \ldots, a_r}_{n_r} \rangle \qquad (n_1 + n_2 + \ldots + n_r = n)$$

SOLUTION: If there were no repetitions, i.e., if all elements were different, the solution would be $n!$, but because of repetitions, the solution is

$$\frac{n!}{n_1!n_2!\ldots n_r!}$$

2.3. Basic Definitions

In the introductory problems we saw that two questions were very important:

- Is there a repetition of objects to be arranged?
- Is their order important?

To emphasize and formalize the importance of these two questions, in this section we define combinations, permutations, etc. Because of the importance of the last few examples in Section 2.2, we repeat them here and later present different derivations of the same formulas.

DEFINITION 2.1 (k-PERMUTATIONS WITHOUT REPETITION). A k-permutation without repetition of the set $A = \{a_1, a_2, \ldots, a_n\}$, $(n \geq k)$ is an arbitrary ordered k-tuple of different elements from that set.

How many different k-permutations of the set A are there? Denote that number as P_n^k. Since $|A| = n$, the first component of the ordered k-tuple can be selected in n different ways. Since the components must be different, the second component can be selected in $(n-1)$ ways, the third in $(n-2)$ ways, etc. Using the product rule, it is now obvious that:

$$P_n^k = n \cdot (n-1) \cdot (n-2) \cdot \ldots \cdot (n-k+1)$$

DEFINITION 2.2 (PERMUTATIONS WITHOUT REPETITION). A permutation without repetition of the set A with n elements is an arbitrary bijection of A onto itself.

It is easy to see that this definition is equivalent to saying that a permutation is just an n-permutation of a set with n elements. Hence, it is uniquely determined by an ordered n-tuple of different elements from A.

If we denote by P_n the number of permutations of the set A with n elements, then:

$$P_n = P_n^n = n \cdot (n-1) \cdot (n-2) \cdot \ldots \cdot (n-n+1)$$

That is,

$$P_n = n!$$

DEFINITION 2.3 (COMBINATIONS WITHOUT REPETITION). A k-combination without repetition of the set A with n elements is an arbitrary subset of A having k elements.

NOTE: The k-combinations are subsets, while k-permutations are ordered k-tuples.

Let C_n^k be the number of all k-combinations of a set with n elements. Since we can form $k!$ different ordered k-tuples from a subset with k elements, we can write

$$P_n^k = C_n^k \cdot k!$$

That is,

$$C_n^k = \frac{P_n^k}{k!} = \frac{n!}{k!\,(n-k)!} = \binom{n}{k}$$

where $\binom{n}{k}$ (read: *n choose k*) is the usual notation:

$$\binom{n}{k} = \begin{cases} n!/(k!\,(n-k)!), & 0 \leq k \leq n \\ 0, & k < 0 \text{ or } k > n. \end{cases}$$

DEFINITION 2.4 (k-PERMUTATIONS WITH REPETITION). A k-permutation with repetition of the set $A = \{a_1, a_2, \ldots, a_n\}$, $(n \geq k)$ is an arbitrary ordered k-tuple of (not necessarily different) elements from that set.

The number of k-permutations with repetition of a set with n elements is denoted by $\overline{P}_n^k$, and it is easy to see that:

$$\overline{P}_n^k = n^k$$

DEFINITION 2.5 (PERMUTATIONS WITH REPETITION). A permutation with repetition of the type $(n_1, \ldots, n_r)$ of the set $A = \{a_1, \ldots, a_r\}$, where $n_1 + \ldots + n_r = n$, is an arbitrary ordered n-tuple, whose components are from A, such that n_1 components equal a_1, n_2 components equal $a_2, \ldots$ and n_r components equal a_r.

The number of permutations with repetition of the type $(n_1, \ldots, n_r)$ is denoted by $\overline{P}_{n_1,\ldots,n_r}$, and it equals (as in Example 2.19):

$$\overline{P}_{n_1,\ldots,n_r} = \frac{n!}{n_1! \ldots n_r!}$$

DEFINITION 2.6 (COMBINATIONS WITH REPETITION). A k-combination with repetition of the set $A = \{a_1, a_2, \ldots, a_n\}$ is an arbitrary k-element multiset of elements from A.

The number of k-combinations with repetition of a set with n elements is denoted by $\overline{C}_n^k$, and it equals (as in Example 2.18):

$$\overline{C}_n^k = \binom{n+k-1}{k}$$

EXAMPLE 2.20. Citizens of a certain town asked their telephone company for a special switchboard. They wanted their phone numbers to begin with 555-6 and one of the following features, whichever is most profitable for the phone company:

1. The last three digits must differ, and all numbers having the same digits at these places, order being irrelevant, should activate the same phone, e.g., 555-6145, 555-6154, ...

2. The last three digits are arbitrary, repetition is allowed. All numbers having the same sum of digits should activate the same phone line, e.g., 555-6249, 555-6555, 555-6366, ...

3. The fifth digit must be 4, while the sixth and the seventh digits are arbitrary. All numbers with the same product of digits must activate the same phone, e.g., 555-6436, 555-6463, 555-6429, ...

Which option was selected by the telephone company if its goal is to achieve the largest capacity, i.e., to have as many telephones as possible in this town?

SOLUTION: In the first case we want to see how many connections the switchboard can have if each connection is determined by three different digits no matter in what order. Let this number be M. If we select three digits from $\{0, 1, 2, \ldots, 9\}$, these three digits determine $3 \cdot 2 \cdot 1 = 6$ different ordered triples.

These three digits can be selected in M different ways; therefore $6M$ is the number of all possible ordered triples composed of different digits. In other words $6M = 10 \cdot 9 \cdot 8 = 720$; i.e., $M = 120$.

As soon as we notice that the order of digits is irrelevant and the digits must be different, we can say these are three-element subsets of the set of digits, i.e., 3-combinations without repetition: $M = \binom{10}{3} = 120$.

In the second case we must know how many different sums can be formed using three digits. On the lower side we have $0+0+0=0$, while on the higher side we have $9+9+9=27$. All numbers from 0–27 can be represented as sums of three digits, so the number of connections in this case is 28. Since $28 < 120$, the first case is better than the second.

In the third case the number of connections is equal to the number of different products of two digits. This number can be found in many different ways, but we can avoid almost any calculations by noting that the number of different products is certainly not greater than 100, because there are exactly 100 ordered couples, $10 \cdot 10 = 100$. It can be shown that there are exactly 37 different products of two digits.

Finally the choice is clear: The type-1 switchboard will be used.

EXAMPLE 2.21. How many diagonals exist in a convex polygon with n sides?

SOLUTION: The points forming the polygon define the total of $\binom{n}{2}$ lines, because every line is defined by a two-element subset of given points. Among these $\binom{n}{2}$ lines, n are the sides of the polygon, so the number of diagonals is

$$\binom{n}{2} - n = \frac{n(n-3)}{2}$$

EXAMPLE 2.22. How many intersections of diagonals of a convex polygon with n sides exist, if the n vertices of the polygon are not counted?

SOLUTION: Two intersecting diagonals must be defined by four different vertices of the polygon or their intersection is one of the vertices. Thus there are as many intersections of the diagonals as there are quadrilaterals formed by the n vertices of the polygon, i.e., $\binom{n}{4}$.

EXAMPLE 2.23. We are given a rectangle divided by horizontal and vertical lines into $m \times n$ squares 1×1. How many different rectangles are defined by this grid?

SOLUTION: Every rectangle is defined by two horizontal and two vertical lines. There are $m+1$ and $n+1$ such lines, respectively. Since the horizontal lines are picked independently from the vertical lines, and the order in which the two horizontal and the two vertical lines are picked is irrelevant, and also repetition is not allowed (otherwise some of the rectangles would have area zero), the solution is $\binom{m+1}{2} \cdot \binom{n+1}{2}$.

EXAMPLE 2.24. How many triples of natural numbers less than 100 have a sum divisible by three if order:

1. Matters and repetition is allowed?
2. Matters and repetition is not allowed?
3. Is irrelevant and repetition is allowed?
4. Is irrelevant and repetition is not allowed?

SOLUTION: We observe first that the sum of three numbers is divisible by 3 if and only if all three numbers are from the same divisibility class or if they all come from different divisibility classes. We also see that each of the three classes has 33 natural numbers less than 100:

1. $3 \cdot 33^3 + 99 \cdot 66 \cdot 33 = 9 \cdot 33^3$.
2. $3 \cdot 33 \cdot 32 \cdot 31 + 99 \cdot 66 \cdot 33 = 3 \cdot 33 \cdot 32 \cdot 31 + 6 \cdot 33^3$.
3. $3 \cdot \binom{33+3-1}{3} + 99 \cdot 66 \cdot 33/3! = 3 \cdot \binom{35}{3} + 33^3$.
4. $3 \cdot \binom{33}{3} + 99 \cdot 66 \cdot 33/3! = 3 \cdot \binom{33}{3} + 33^3$.

EXAMPLE 2.25. How many positions of eight rooks are there on a chess board in which no two of them attack each other?

SOLUTION: First of all each row and column may contain only one rook. In Column A we have eight possibilities. After making a choice there, in Column B we have seven possibilities, etc. At the end for Column H we have no other choice but to put the last rook in the only remaining position. The total number of positions is therefore $8 \cdot 7 \cdot \ldots \cdot 1 = 8! = 40320$.

EXAMPLE 2.26. How many paths are there for the king from A-1 to H-8 if it moves only forward, right, or forward-right?

SOLUTION: Consider first a lame king, which cannot make diagonal moves. No matter which path he takes from A-1 to H-8, he always makes 14 moves, seven forward and seven to the right. Paths differ only by the order of the forward and right moves. Hence the solution for the lame king is

$$\frac{14!}{7!7!} = 3432$$

Since this number also equals $\binom{14}{7}$, this part of the problem can be solved by counting the number of ways of selecting seven forward moves out of 14.

The problem with the healthy king is more complicated because he can additionally move diagonally forward-right. If the king makes k diagonal moves, then the number of forward and right moves is $7-k$ each. Then for each $k = 0, 1, \ldots, 7$ the number of paths is

$$\frac{(k+(7-k)+(7-k))!}{k!\,(7-k)!\,(7-k)!} = \frac{(14-k)!}{k!\,(7-k)!\,(7-k)!}$$

Therefore the total solution is

$$\sum_{k=0}^{7} \frac{(14-k)!}{k!\,(7-k)!\,(7-k)!} = 48639$$

EXAMPLE 2.27 (BINOMIAL COEFFICIENTS). Prove the following identities:

1. Newton's binomial formula:

$$(a+b)^n = \sum_{k=0}^{n} \binom{n}{k} a^{n-k} b^k =$$
$$= \binom{n}{0} a^n b^0 + \binom{n}{1} a^{n-1} b^1 + \ldots + \binom{n}{n} a^0 b^n$$

2. The sum of the binomial coefficients:

$$\sum_{k=0}^{n} \binom{n}{k} = \binom{n}{0} + \binom{n}{1} + \ldots + \binom{n}{n} = 2^n$$

3. Symmetry of binomial coefficients:

$$\binom{n}{k} = \binom{n}{n-k}$$

4. Pascal's formula:

$$\binom{n}{k} = \binom{n-1}{k-1} + \binom{n-1}{k}$$

SOLUTION OF 1: Earlier we proved that the number of k-element subsets of an n-element set is

$$C_n^k = \frac{n!}{k!\,(n-k)!}$$

We also mentioned that instead of the clumsy quotient of factorials, it is more convenient to write

$$C_n^k = \binom{n}{k} \qquad \binom{n}{k} = \begin{cases} n!/(k!\,(n-k)!), & 0 \le k \le n \\ 0, & k < 0 \text{ or } k > n \end{cases}$$

and that the numbers $\binom{n}{k}$ are called binomial coefficients. Now we see where this name came from.

If we calculate $\binom{n}{k}$ for the first several values of n and k, we can form a table in Fig. 2.2:

We obtain the same numbers that appear in the fully expanded powers of a binomial:

$$\begin{aligned}
(a+b)^0 &= 1 \\
(a+b)^1 &= a+b \\
(a+b)^2 &= a^2+2ab+b^2 \\
(a+b)^3 &= a^3+3a^2b+3ab^2+b^3 \\
(a+b)^4 &= a^4+4a^3b+6a^2b^2+4ab^3+b^4 \\
&\vdots
\end{aligned}$$

$n \backslash k$	0	1	2	3	4
0	1	0	0	0	0
1	1	1	0	0	0
2	1	2	1	0	0
3	1	3	3	1	0
4	1	4	6	4	1

FIGURE 2.2. The first several values of C_n^k, i.e., binomial coefficients (Pascal's triangle).

In general Newton's binomial formula holds

$$(a+b)^n = \sum_{k=0}^{n} \binom{n}{k} a^{n-k} b^k$$

This can be proved by mathematical induction (see Problem A.8, Appendix A) but also by using the following combinatorial thinking.

Expanding the expression $(a+b)^n$ yields a sum whose terms all have the form $A_k a^{n-k} b^k$, $0 \leq k \leq n$, where the numbers A_k are called the binomial coefficients. We next show that A_k equals the number of k-combinations of an n-element set.

In the equality:

$$\underbrace{(a+b)(a+b)\ldots(a+b)}_{n} = \sum_{k=0}^{n} A_k a^{n-k} b^k$$

the term $a^{n-k} b^k$ appears as many times as there are ways of selecting k letters b from n boxes. Also the order of boxes is not important because the order is irrelevant in multiplication; it always yields b^k. Therefore the binomial coefficients equal the number of k-combinations of an n-element set. In other words:

$$A_k = \binom{n}{k}$$

which justifies the name binomial coefficients.

SOLUTION OF 2: In Newton's binomial formula set $a = b = 1$ to obtain the desired identity. Here is another, combinatorial proof: In Example 2.14 the total number of subsets of an n-element set is 2^n. We can enumerate them by counting all k-element subsets for each $k = 0, 1, 2, \ldots, n$, so we can write

$$\sum_{k=0}^{n} \binom{n}{k} = \binom{n}{0} + \binom{n}{1} + \ldots + \binom{n}{n} = 2^n$$

SOLUTION OF 3: The symmetry of the binomial coefficients can be proved by algebraically manipulating the formulas. Here we prove it in a combinatorial manner: Every k-element subset is determined by the k elements it contains

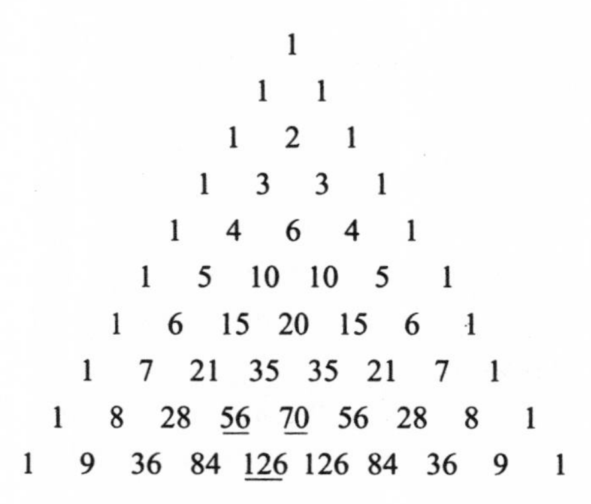

FIGURE 2.3. Pascal's triangle. According to Pascal's formula 126 was obtained as $56+70$.

but also by the $(n-k)$ elements it does not contain (see also Example 1.35). Therefore:

$$\binom{n}{k} = \binom{n}{n-k}$$

SOLUTION OF 4: Pascal's formula, too, can be proved both algebraically and combinatorially; see also Example 2.35. The combinatorial proof follows.

Among n elements of the initial set A we select one and call it x. All k-element subsets of A are divided into two disjoint groups according to whether or not these contain x. There are $\binom{n-1}{k-1}$ subsets containing x because besides x, we are free to pick $(k-1)$ elements from the $(n-1)$-element set $A-\{x\}$. Similarly there are $\binom{n-1}{k}$ subsets not containing x; hence:

$$\binom{n}{k} = \binom{n-1}{k-1} + \binom{n-1}{k}$$

Using Pascal's formula and the symmetry of the binomial coefficients, we can form a larger Pascal's triangle (Fig. 2.3):

NOTES: Let us mention a few facts from the history of Pascal's triangle and the binomial theorem. In Euclid's *Elements* we find the case of the binomial theorem for $n=2$. Pascal's triangle was known to Chinese, Hindu, and Arab mathematicians in the thirteenth century. The term binomial coefficients was first used by Stifel in the sixteenth century. He showed how to find the expansion of $(1+x)^n$ if we know the expansion of $(1+x)^{n-1}$ (Pascal's formula).

The connection between Pascal's triangle and the binomial coefficients was first discovered by Tartaglia in the sixteenth century. In the seventeenth century, Pascal published his treatise on binomial coefficients, which showed the connection between combinations and binomial coefficients. Newton was the first to consider the rational powers of binomials. He also found an efficient way of expanding $(1+x)^n$ without prior knowledge of $(1+x)^{n-1}$. His method is based on the formula:

$$\binom{n}{k+1} = \frac{n-k}{k+1}\binom{n}{k}$$

We leave to the reader the pleasure of proving this formula and discovering how to apply it.

EXAMPLE 2.28. How many ways are there of giving 30 books to seven friends?

SOLUTION: We observe that this question is not precise because it does not specify whether the books are the same, different, or perhaps five copies of one title, six of another, and 19 of yet another.

In addition if the books are not all the same, it may matter whether a person is given books in a particular order.

Another possibility is that we wish to make seven gift packages, so that who receives which package does not matter.

Let us consider a few possibilities:

1. Let the books differ. Each of the 30 books can go to one of seven addresses, so the total number of possibilities is 7^{30}. Obviously these are 30-permutations of the set of seven friends, but the solution is simple enough not to make a mention of permutations.

2. Let us assume again that all the books differ. In the previous case we considered as solutions even arrangements where some friends did not receive books. What if we want to make sure each of seven friends receives at least one book? Let us first give one book to each friend. This can be done in $30 \cdot 29 \cdot 28 \cdot 27 \cdot 26 \cdot 25 \cdot 24$ ways. The remaining 23 books can then be given in 7^{23} ways, making the total:

$$30 \cdot 29 \cdot 28 \cdot 27 \cdot 26 \cdot 25 \cdot 24 \cdot 7^{23}$$

3. Now assume the books are all the same. On paper draw six vertical lines. To the left of the first line draw two circles to represent that

the first friend receives two copies. Draw five circles between the first and the second line to represent five copies given to the second friend. Continue like that, then finally draw one circle to the right of the last line for the one copy given to the seventh friend. There must be 30 circles. Every such arrangement of six lines and 30 circles uniquely represents one arrangement of the books. Note: Every book arrangement can be uniquely represented by one such diagram. The number of diagrams with lines and circles is $36!/(30!\,6!) = \binom{36}{30} = \binom{36}{6}$.

4. Let the books be the same but make sure that each friend receives at least one book. This case can be reduced to the previous one in the following way:

   Case 3 is equivalent to the problem of counting all different solutions of the equation:

$$x_1 + x_2 + \ldots + x_7 = 30 \qquad x_i \in N_0 \quad (i = 1, 2, \ldots, 7)$$

   if solutions such as $(30, 0, 0, \ldots, 0)$ and $(0, 30, 0, \ldots, 0)$ are considered as different. (The first of them gives all books to the first friend, while the second gives all books to the second friend.) As we saw earlier, the number of such solutions is $\binom{36}{6}$.

   In Case 4 we are looking for strictly positive solutions because each friend must receive at least one book:

$$y_1 + y_2 + \ldots + y_7 = 30 \qquad y_i \in N \quad (i = 1, 2, \ldots, 7)$$

   We subtract 7 from both sides to reduce the problem to one similar to that in 3:

$$(y_1 - 1) + (y_2 - 1) + \ldots + (y_7 - 1) = 30 - 7$$
$$z_1 + z_2 + \ldots + z_7 = 23 \quad z_i \in N_0 \quad z_i = y_i - 1 \quad (i = 1, 2, \ldots, 7)$$

   The number of solutions of the preceding equation, i.e., the number of arrangements of 30 books to seven friends, such that every friend receives at least one book is therefore $\binom{29}{23} = \binom{29}{6}$. □

Among other things in Example 2.27 we saw that the equation:

$$x_1 + x_2 + \ldots + x_n = k \qquad x_i \in N_0 \quad (i = 1, 2, \ldots, n)$$

has $\binom{n+k-1}{k}$ solutions. Among them some solutions can be considered equal, e.g., $(k, 0, 0, \ldots, 0)$ and $(0, k, 0, \ldots, 0)$, but in Example 2.27 we wanted to count them as different.

Although Euler's famous problem *partitio numerorum* is similar to Example 2.27, it is much more difficult to solve:

The number 4 can be written as a sum of *one or more* natural numbers, where the order of the terms is *irrelevant*, in five different ways:

$$4 \qquad 1+3 \qquad 2+2 \qquad 1+1+2 \qquad 1+1+1+1$$

We say that 4 has five *partitions* and write $p(4) = 5$. Later we discuss several theorems about different types of partitions. For the time being we just mention that the expression for $p(n)$ was determined by Rademacher in 1934. His complicated formula can be substituted with the following asymptotic formula, found by Hardy and Ramanujan in 1917:

$$p(n) \sim \frac{1}{4n\sqrt{3}} e^{\pi\sqrt{\frac{2}{3}n}}$$

We mention here a much more important asymptotic formula, the famous Stirling approximation from 1730 (actually discovered by de Moivre):

$$n! \sim \sqrt{2\pi n} \left(\frac{n}{e}\right)^n$$

EXAMPLE 2.29 (SYLVESTER'S FORMULA). Generalize the inclusion–exclusion principle to n sets $A_1, \ldots, A_n$.

RESULT: Use mathematical induction to prove that if:

$$\begin{aligned}
S_1 &= |A_1| + |A_2| + \ldots + |A_n| \\
S_2 &= |A_1 \cap A_2| + |A_1 \cap A_3| + \ldots + |A_{n-1} \cap A_n| \\
&\vdots \\
S_n &= |A_1 \cap A_2 \cap \ldots \cap A_n|
\end{aligned}$$

then:

$$|A_1 \cup A_2 \cup \ldots \cup A_n| = S_1 - S_2 + S_3 - S_4 + \ldots + (-1)^{n+1} S_n = \sum_{k=1}^{n} (-1)^{k+1} S_k$$

This formula was also first discovered by de Moivre, but it bears Sylvester's name because he often used it.

EXAMPLE 2.30 (EULER'S PHI FUNCTION). Let $p_1, \ldots, p_r$ be the prime factors of the integer $n > 1$. Find $\varphi(n)$, the number of numbers less than n and relatively prime to it.

SOLUTION: Euler's function $\varphi(n)$ is very important in the theory of numbers. For example Euler's theorem states:

$$\text{If } (a,m) = 1, \text{ then } a^{\varphi(n)} \equiv 1 \pmod{n}$$

We prove that if the canonical decomposition of n is given by $n = p_1^{\alpha_1} \ldots p_r^{\alpha_r}$, then:

$$\varphi(n) = n\left(1 - \frac{1}{p_1}\right) \ldots \left(1 - \frac{1}{p_r}\right)$$

Let A_i be the set of all numbers $\leq n$ divisible by p_i $(i = 1, \ldots, r)$. Then the union of all sets A_i is the set of all numbers not relatively prime to n, hence:

$$\varphi(n) = n - s$$

where $s = |A_1 \cup A_2 \cup \ldots \cup A_n|$.

Since $|A_i| = n/p_i$ and $|A_i \cap A_j| = n/(p_i p_j)$, etc., according to the inclusion–exclusion principle, we have

$$\begin{aligned} \varphi(n) &= n - s \\ &= n - \left[\frac{n}{p_1} + \ldots + \frac{n}{p_r} - \left(\frac{n}{p_1 p_2} + \ldots + \frac{n}{p_{r-1} p_r}\right) + \ldots + (-1)^{r+1} \frac{n}{p_1 \ldots p_r}\right] \end{aligned} \quad (2.1)$$

It is easy to check that the last expression equals

$$\varphi(n) = n\left(1 - \frac{1}{p_1}\right) \ldots \left(1 - \frac{1}{p_r}\right)$$

EXAMPLE 2.31 (DERANGEMENTS). How many permutations of $\{1,2,\ldots,n\}$ are such that k is not at the kth place for any k $(1 \leq k \leq n)$? Such permutations are called *derangements*.

SOLUTION: There are many formulations of this problem, for example the Bernoulli*–Euler problem of misaddressed letters: How many ways can a math professor incorrectly address Christmas cards so that no card gets to the originally intended recipient?

If from the total number of permutations, i.e., from $n!$, we subtract the number of permutations in which at least one element k is in the kth $(k = 1,2,\ldots,n)$ place [there are $n \cdot (n-1)!$ such permutations], we have

$$n! - n!$$

We see that we overdid it. For example the permutation in which numbers 1 and 2 are in the first and the second places, respectively, and all other elements are deranged, was subtracted twice, once because of 1 and once because of 2.

To account for this, we must return the number of permutations in which two or more elements are in forbidden places, a total of $\binom{n}{2}(n-2)!$ permutations:

$$n! - n! + \binom{n}{2}(n-2)!$$

Continuing this correction process, we finally obtain the number of derangements of an n-element set:

$$D_n = n!\left[1 - 1 + \frac{1}{2!} - \frac{1}{3!} + \ldots + (-1)^n \frac{1}{n!}\right]$$

Since D_n is an integer and

$$1 - 1 + \frac{1}{2!} - \frac{1}{3!} + \ldots + (-1)^n \frac{1}{n!} + \ldots = \frac{1}{e}$$

The D_n is the closest integer to the number $n!/e$, i.e.:

$$D_n = \left\lfloor \frac{n!}{e} + \frac{1}{2} \right\rfloor$$

*We must specify that it was Nicolaus (I) Bernoulli, because this Swiss family produced an enormous number of important mathematicians and scientists, among others Jakob, Johann, Daniel, Nicolaus (I), and Nicolaus (II).

where $\lfloor x \rfloor$ is the first integer $\leq x$.

We see that the probability of having all cards sent incorrectly is rather large:

$$\frac{D_n}{n!} \approx \frac{1}{e} = 0.36792\ldots$$

The recursion for the number of derangements is $D_n = (n-1)(D_{n-1} + D_{n-2})$, with $D_1 = 0$ and $D_2 = 1$. Other initial conditions give other solutions. For example using 1 and 2 as initial conditions yields $n!$ as a solution. □

Fibonacci, better known among his contemporaries as Leonardo of Pisa or Leonardo Pisano, is remembered today mostly for the sequence of numbers appearing as a solution to his problem about rabbits. He published that problem in his book *Liber Abaci* in 1202 (see Example 2.32). His contribution to Western civilization however is much greater, for he insisted that Hindu–Arabic numerals be used instead of Roman numerals. The change was difficult, but succeeded because of the many advantages of Hindu–Arabic numerals in calculating and accounting.

EXAMPLE 2.32 (FIBONACCI NUMBERS). Rabbits mature one month after birth. Each month a mature pair of rabbits gives birth to a new pair of rabbits. If we begin with a newly born pair of rabbits, how many pairs do we have at the beginning of the nth month? Solve the problem as if no rabbits die during these n months.

SOLUTION: If at the beginning of the $(n-1)$st month there are f_{n-1} pairs, then at the beginning of the next, nth month we have all the pairs we had at the beginning of $(n-1)$st month, i.e., f_{n-1}, plus the babies of the pairs we had at the beginning of the $(n-2)$nd month, i.e., f_{n-2}. Then:

$$f_n = f_{n-1} + f_{n-2} \qquad f_1 = 1 \quad f_2 = 1$$

Using this recursive relation we can find that for $n = 1, 2, 3, 4, 5, 6, 7, 8, 9, \ldots$

$$f_1 = 1 \; f_2 = 1 \; f_3 = 2 \; f_4 = 3 \; f_5 = 5 \; f_6 = 8 \; f_7 = 13 \; f_8 = 21 \; f_9 = 34 \; \ldots$$

Usually we write $f_0 = 0$.

NOTES: Fibonacci numbers posses a number of properties, which we investigate to some extent here and in Appendix A. The recursion $f_{n+1} = f_n + f_{n-1}$ was first used by Girard in 1634. Simson noted in 1753 that as n increases, the ratio f_{n+1}/f_n converges to the golden section:

$$\phi = \frac{1+\sqrt{5}}{2} = 1.6180\ldots$$

This ubiquitous sequence of numbers was given this name only in the nineteenth century by the French mathematician Lucas.

Fibonacci numbers are often encountered in mathematical problems of various kinds and in nature too. For example in a row of seeds in a sunflower head, there is a Fibonacci number of seeds, e.g., 55, even more but always one of the numbers f_n!! □

Before we go on to generating functions, recall that Newton's binomial formula can be generalized using the Maclaurin series for $(1+x)^\alpha$. According to Maclaurin's formula and Abel's convergence criterion, we find

$$(1+x)^\alpha = 1 + \alpha x + \frac{\alpha(\alpha-1)}{2}x^2 + \ldots + \frac{\alpha(\alpha-1)\ldots(\alpha-k+1)}{k!}x^k + \ldots$$

$$(|x| < 1,\ \alpha \in R)$$

When $\alpha = n$, this expression has a finite number of terms, and it reduces to Newton's binomial formula.

2.4. Generating Functions

Besides the algebraic and combinatorial methods of proofs, there is the third general method used in enumeration and to prove identities and properties of binomial coefficients — the method of generating functions. Since many other problems in mathematics can be solved by generating functions, we briefly introduce this method, first used by de Moivre and later improved by Euler and Laplace.

EXAMPLE 2.33. Consider the identity $(1+x)^n(1+x)^n = (1+x)^{2n}$. Coefficients next to x^n on the left- and right-hand side of the equality sign must be equal, so that:

$$\binom{n}{0}\binom{n}{n} + \binom{n}{1}\binom{n}{n-1} + \ldots + \binom{n}{n}\binom{n}{0} = \binom{2n}{n}$$

Since the binomial coefficients are symmetric, i.e., $\binom{n}{k} = \binom{n}{n-k}$, we obtain the identity:

$$\binom{n}{0}\binom{n}{0} + \binom{n}{1}\binom{n}{1} + \ldots + \binom{n}{n}\binom{n}{n} = \binom{2n}{n}$$

That is,

$$\binom{n}{0}^2 + \binom{n}{1}^2 + \ldots + \binom{n}{n}^2 = \binom{2n}{n}$$

Here we use the fact that in a power of a binomial all binomial coefficients are hidden. We say that $(1+x)^n$ generates the binomial coefficients or that it is the generating function of the binomial coefficients. But before defining this new terminology, let us consider a few more examples.

EXAMPLE 2.34. If we multiply three binomials:

$$(1+ax)(1+bx)(1+cx) = 1 + (a+b+c)x + (ab+ac+bc)x^2 + abcx^3$$

next to x^k $(k = 0,1,2,3)$ we have all k-combinations of the set $\{a,b,c\}$.

Similarly in the product:

$$\begin{aligned}&(1+a_1x)(1+a_2x)\ldots(1+a_nx) = \\ &= 1 + (a_1 + \ldots + a_n)x + (a_1a_2 + a_1a_3 + \ldots + a_{n-1}a_n)x^2 + \ldots + a_1\ldots a_nx^n\end{aligned} \tag{2.2}$$

we find a list of all k-combinations of $\{a_1,a_2,\ldots,a_n\}$ next to x^k $(k = 0,1,\ldots,n)$.

To count them, we just set $a_1 = a_2 = \ldots = a_n = 1$ and next to x^k we can find the number of k-combinations of an n-element set, a fact that we found earlier from Newton's binomial formula.

Thus we see that the function $C_n(x) = (1+x)^n$ generates binomial coefficients $\binom{n}{0}, \binom{n}{1}, \ldots, \binom{n}{n}$; therefore it is called the generating function of binomial coefficients.

EXAMPLE 2.35 (AGAIN PASCAL'S FORMULA). From $C_n(x) = (1+x)^n$ it follows that:

$$C_n(x) = (1+x)C_{n-1}(x) = C_{n-1}(x) + xC_{n-1}(x)$$

Equating coefficients next to x^k on the two sides, we obtain Pascal's formula:

$$\binom{n}{k} = \binom{n-1}{k} + \binom{n-1}{k-1}$$

EXAMPLE 2.36. What happens if we multiply $(1+ax+a^2x^2)(1+bx)(1+cx)$?

$$\begin{aligned}&(1+ax+a^2x^2)(1+bx)(1+cx) = \\ &= 1+(a+b+c)x+(a^2+ab+ac+bc)x^2+(a^2b+a^2c+abc)x^3+a^2bcx^4\end{aligned} \tag{2.3}$$

Coefficients next to x^k list combinations neither with nor without repetition in some kind of hybrid combinations in which the element a can be repeated at most twice. To enumerate how many such combinations there are, it suffices to set $a = b = c = 1$, then to examine the coefficient next to the corresponding power x^k.

EXAMPLE 2.37 (AGAIN COMBINATIONS WITH REPETITIONS). From the previous example we can learn that to construct a generating function for combinations with repetitions, we must allow every element to appear an arbitrary number of times. Hence the generating function for combinations with repetition of the set $\{a,b,c\}$ is

$$(1+ax+a^2x^2+\ldots)(1+bx+b^2x^2+\ldots)(1+cx+c^2x^2+\ldots)$$

To see their numbers, set $a = b = c = 1$ and observe the corresponding coefficients.

In general the generating function for the number of k-combinations with repetition of an n-element set is

$$\overline{C}_n(x) = (1+x+x^2+\ldots)^n = \frac{1}{(1-x)^n} = (1-x)^{-n}$$

As we mentioned earlier, according to Maclaurin's formula, we have

$$(1-x)^{-n} = \sum_{k=0}^{\infty}(-1)^k \frac{(-n)(-n-1)\ldots(-n-(k-1))}{k!}x^k$$

That is,

$$\overline{C}_n(x) = \sum_{k=0}^{\infty}\binom{n+k-1}{k}x^k$$

We could have expected this result from our earlier derivations.

EXAMPLE 2.38. Let us try a small experiment again. What kinds of problems can be solved by the following generating function:

$$(1+x+x^2+x^3+x^4+x^5+x^6+\dots)(1+x^2+x^4+x^6+x^8+x^{10}+x^{12}+\dots)$$
$$=1+x+2x^2+2x^3+3x^4+3x^5+4x^6+\dots$$

The fact that the coefficient next to x^5 equals 3, tells us that x^5 was produced in three different ways, namely:

$$x^5 \cdot 1 + x^3x^2 + xx^4$$

This reminds us that number 5 can be represented by using only numbers 1 and 2 in exactly three ways:

$$1+1+1+1+1 \qquad 1+1+1+2 \qquad 1+2+2$$ □

Example 2.38 suggests a way of determining the generating function for the number of partitions of the integer n. Recall that the partition of n is any particular way of writing n as a sum of one or more positive integers. It is important to emphasize that the order of terms is irrelevant, for example $1+3+4$ and $1+4+3$ represent the same partition of 8.

The generating function of the sequence $p(n)$ is

$$P(x) = (1+x+x^2+\dots)(1+x^2+x^4+\dots)(1+x^3+x^6+\dots)\times$$
$$\times(1+x^4+x^8+\dots)\dots$$

That is,

$$P(x) = \frac{1}{1-x}\frac{1}{1-x^2}\frac{1}{1-x^3}\frac{1}{1-x^4}\cdots$$

By expanding this formula and using number theory, Hardy and Ramanujan derived their approximation.

Let us take a look at some other types of partitions.

EXAMPLE 2.39. The generating function for partitions using numbers a, b, c, and d, such that a may be used an arbitrary number of times, b can be used at most three times, c must be used at least twice, while d must be used an even number of times is

$$(1+x^a+x^{2a}+\dots)(1+x^b+x^{2b}+x^{3b})(x^{2c}+x^{3c}+\dots)(1+x^{2d}+x^{4d}+\dots)$$

EXAMPLE 2.40 (CHANGE). The number of ways of making change for 1 dinar using an arbitrary number of coins of 1, 2, 5, 10, 20, and 50 paras can be calculated as the coefficient next to x^{100} in the expansion of the following generating function:

$$\begin{gathered}(1+x+\dots)(1+x^2+\dots)(1+x^5+\dots)\times\\ \times(1+x^{10}+\dots)(1+x^{20}+\dots)(1+x^{50}+\dots)\end{gathered}$$

Using polynomial multiplication, the computer gives us the solution of 4562 ways of making change for 1 dinar.

EXAMPLE 2.41 (EULER'S THEOREM ABOUT PARTITIONS). We prove that the number of partitions using different natural numbers, $p_r(n)$, equals the number of partitions using (not necessarily different) odd numbers, $p_n(n)$. The proof shows that two sequences $\{p_r(n)\}$ and $\{p_n(n)\}$ have the same generating functions. Indeed:

$$P_r(x)=(1+x)(1+x^2)(1+x^3)(1+x^4)\dots$$

while

$$P_n(x)=\frac{1}{(1-x)}\frac{1}{(1-x^3)}\frac{1}{(1-x^5)}\frac{1}{(1-x^7)}\cdots$$

Since

$$1+x^k=\frac{1-x^{2k}}{1-x^k}\qquad(k=1,2,\dots)$$

after appropriate cancellations, we find $P_r(x)=P_n(x)$, i.e., $p_r(n)=p_n(n)$. □

Definition 2.7 formally defines generating functions.

DEFINITION 2.7. Generating function of the sequence $(a_0, a_1, \ldots, a_k, \ldots)$ is the power series:

$$A(x) = a_0 + a_1 x + \ldots + a_k x^k + \ldots$$

EXAMPLE 2.42. Examples follow of some sequences and their generating functions:

$$\begin{aligned}
(1,0,0,1,0,0,\ldots) &\leftrightarrow 1 + x^3 \\
(1,1,1,1,1,1,\ldots) &\leftrightarrow 1 + x + x^2 + x^3 + \ldots = \frac{1}{1-x} \\
(1,0,1,0,1,0,\ldots) &\leftrightarrow 1 + x^2 + x^4 + x^6 + \ldots = \frac{1}{1-x^2} \\
(1,2,3,4,5,6,\ldots) &\leftrightarrow 1 + 2x + 3x^2 + 4x^3 + \ldots = \frac{1}{(1-x)^2} \\
(0,1,2,3,4,5,\ldots) &\leftrightarrow x + 2x^2 + 3x^3 + 4x^4 + \ldots = \frac{x}{(1-x)^2} \\
\left(1, 1, \frac{1}{2!}, \frac{1}{3!}, \frac{1}{4!}, \ldots\right) &\leftrightarrow 1 + x + \frac{x^2}{2!} + \frac{x^3}{3!} + \frac{x^4}{4!} + \ldots = e^x \\
\left(\binom{n}{0}, \binom{n}{1}, \binom{n}{2}, \ldots, \binom{n}{n}\right) &\leftrightarrow (1+x)^n
\end{aligned}$$

□

It is often useful to form the generating function of a sequence to see what can be accomplished through its transformations. Examples 2.43 and 2.44 use this approach to investigate Fibonacci numbers.

EXAMPLE 2.43. Generating function for the sequence of Fibonacci numbers is

$$F(x) = f_0 + f_1 x + f_2 x^2 + \ldots$$

Since $xF(x) = f_0 x + f_1 x^2 + f_2 x^3 + \ldots$ and $x^2 F(x) = f_0 x^2 + f_1 x^3 + f_2 x^4 + \ldots$ while $f_0 = 0, f_1 = 1$, and for $n \geq 2$ we have $f_n = f_{n-1} + f_{n-2}$, we find that:

$$F(x) - xF(x) - x^2 F(x) = x$$

That is,

$$F(x) = \frac{x}{1 - x - x^2}$$

Reciprocals of roots of the trinomial in the denominator of $F(x)$ are ϕ (the golden section) and $\hat{\phi} = -1/\phi$:

$$\phi = \frac{1+\sqrt{5}}{2} \qquad \hat{\phi} = \frac{1-\sqrt{5}}{2}$$

If $F(x)$ is written in the form of partial fractions, we find that:

$$F(x) = \frac{\sqrt{5}}{5}\frac{1}{1-\phi x} - \frac{\sqrt{5}}{5}\frac{1}{1-\hat{\phi} x}$$

That is,

$$F(x) = \frac{\sqrt{5}}{5}(1+\phi x+\phi^2 x^2+\ldots) - \frac{\sqrt{5}}{5}(1+\hat{\phi} x+\hat{\phi}^2 x^2+\ldots)$$

which finally yields

$$F(x) = 0 + \frac{\sqrt{5}}{5}(\phi - \hat{\phi})x + \frac{\sqrt{5}}{5}(\phi^2 - \hat{\phi}^2)x^2 + \ldots$$

This implies

$$f_n = \frac{\sqrt{5}}{5}(\phi^n - \hat{\phi}^n)$$

That is,

$$f_n = \frac{\sqrt{5}}{5}\left[\left(\frac{1+\sqrt{5}}{2}\right)^n - \left(\frac{1-\sqrt{5}}{2}\right)^n\right]$$

NOTES: This formula was derived by de Moivre in 1718. Historically this was the first known use of generating functions. A different proof was given by Nicolaus (I) Bernoulli in 1728, and it was first published by Euler in 1765. The proof was later forgotten, then rediscovered by Binet in 1843. Today it is called *Binet's formula*.

The number $\phi = 1.61803\ldots$ is called the golden section, and it is also denoted as g or τ. The notation ϕ is used in honor of Phidias, the Ancient Greek sculptor, who believed that objects having proportions dominated by the golden section were most pleasing to the eye. The number ϕ, just like Fibonacci numbers, is encountered very often, not only in mathematics but also in the natural sciences and the arts. See Appendix B for more about this and some other important numbers.

Since Fibonacci numbers are integers, and for $n \geq 0$ we have $\left|\frac{\sqrt{5}}{5}\hat{\phi}^n\right| < \frac{1}{2}$, we see that f_n is the integer closest to $\frac{\sqrt{5}}{5}\phi^n$, i.e.,

$$f_n = \left\lfloor \frac{\sqrt{5}}{5}\phi^n + \frac{1}{2} \right\rfloor \qquad (n \geq 0)$$

EXAMPLE 2.44. If we write $F(x)$ in the following form:

$$F(x) = \frac{x}{1-x-x^2} = \frac{x}{1-x(1+x)} = x\{1+[x(1+x)]+[x(1+x)]^2+\ldots\}$$

we easily find that:

$$\sum_{k=0}^{n}\binom{n-k}{k} = \binom{n}{0}+\binom{n-1}{1}+\binom{n-2}{2}+\ldots = f_{n+1}$$

For example:

$$\binom{6}{0}+\binom{5}{1}+\binom{4}{2}+\binom{3}{3} = 1+5+6+1 = 13 = f_7$$ □

Generating functions have many other properties and applications, especially in probability theory, solving recursions, and such engineering disciplines, as communications and digital signal processing, where they are called the z-transform.

2.5. Problems

EXAMPLE 2.45. There are 100 competitors in a tennis tournament. The competition is organized as a cup, i.e., the competitor who looses a match must leave the tournament. How many matches must be played before we know the winner of the tournament?

SOLUTION: Instead of considering all possible ways of organizing matches among competitors, it suffices to note that every match eliminates one player.

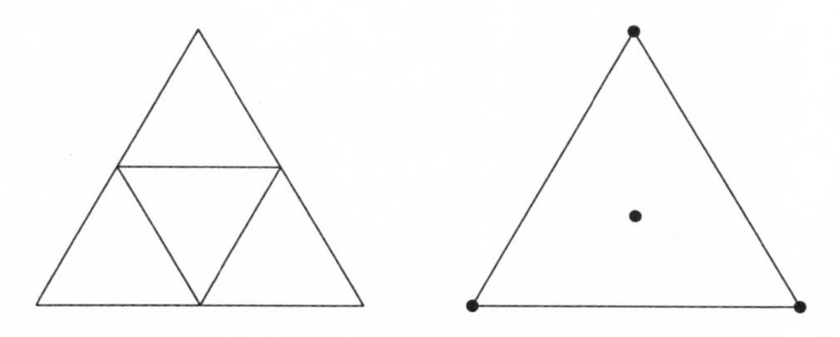

FIGURE 2.4. Examples from combinatorial geometry.

At the end of the tournament, only one player remains unbeaten while each of the remaining 99 players lost one match. The number of matches, which equals the number of losers, is therefore 99.

EXAMPLE 2.46. Mark five arbitrary points in the interior of an equilateral triangle with sides $a = 1$, then show that no matter which five points are chosen, two of them are at a distance $\leq 1/2$.

SOLUTION: Divide the given triangle in four smaller triangles (see Fig. 2.4), then according to Dirichlet's principle, at least one of these four triangles contains two marked points. Note: The points inside a small triangle are all at distances $\leq 1/2$.

EXAMPLE 2.47. An equilateral triangle with $a = 1$ cannot be covered by three circles whose diameters are $d < 1/\sqrt{3}$. Prove.

SOLUTION: No matter how we place three circles of diameters $< 1/\sqrt{3}$, each of them can at most cover one of the following four points: three vertices and the center of the triangle (see Fig. 2.4). According to Dirichlet's principle, any arrangement of the three circles leaves one of the points uncovered.

EXAMPLE 2.48. Is it possible to draw the diagram in Fig. 2.5 without lifting a pencil from the paper and without doubling any of the lines?

SOLUTION: Note that the number of lines connected at each of the vertices of the square is three, i.e., an odd number. Hence to draw the entire diagram in one move, the number of times the pencil enters and exits each vertex must be odd but this is possible only for the vertex where we start the drawing and for the vertex where we end it and only when the starting and the ending vertices are different. Thus we can account for only two odd-degree vertices, not for

FIGURE 2.5. Simplest example of Euler's theorem.

all four of them. This means that drawing in one move is impossible for the given diagram.

NOTES: In a similar manner, in 1736 Euler answered the question that troubled residents Königsberg (today Kaliningrad, Russia): Is it possible to cross the bridges on Pregel River (today, Pregolya) so that each bridge is crossed exactly once? At that time the city had seven bridges connecting two islands and the two banks. The diagram representing bridges, islands, banks, and the river is a graph with four odd-degree vertices (see Fig. 2.6), so reasoning as we did before, we find that residents of Königsberg could not find the desired path no matter how hard they tried.

In graph theory there is a theorem due to Euler, inspired by this famous problem. This theorem and the year 1736 are considered the beginning of graph theory.

EXAMPLE 2.49. How many knights can be arranged on a chess board so that they do not attack each other?

SOLUTION: Since a knight on a white field attacks only black fields, we can place 32 knights on a chess board without any of them attacking each other. On the other hand, placing 33 knights is impossible because in that case some rectangle of 2×4 fields contains five or more knights, which cannot be achieved.

NOTE: In 1850 Nauck showed that there are exactly 92 different peaceful

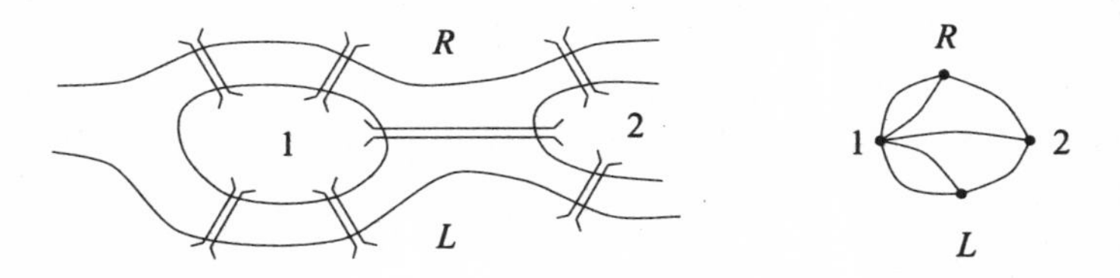

FIGURE 2.6. Königsberg bridges.

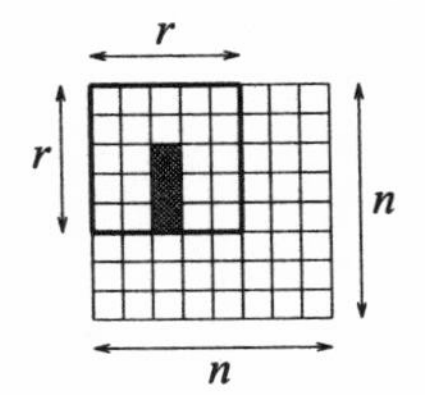

FIGURE 2.7. Rectangles in a square grid.

positions for eight queens on a chess board.

EXAMPLE 2.50. If each domino covers two neighboring fields of an $n \times n$ chess board, prove that a complete covering is possible if and only if n is an even number.

SOLUTION: If n is even, it is easy to see that the covering is possible. If n is odd, the number of fields (n^2) is also odd. Since each domino covers two fields, there is no arrangement that can cover an odd number of fields.

EXAMPLE 2.51. A 6×6 table can be covered by 18 dominoes. Prove that an arbitrary covering of the table can be divided by a straight line so that none of the dominoes is cut.

SOLUTION: If the table is 6×6, there are five possible dividing lines in each horizontal and vertical direction. All these lines cut an even number of dominoes. Indeed if a line cuts an odd number of dominoes, the table will be divided into two parts, each containing an odd number of domino halves. But that implies each of these two parts has an odd number of fields covered by whole dominoes. This is impossible (see Example 2.50). Thus every line cuts an even number of dominoes, so all arrangements of 18 dominoes on ten lines are such that at least one line cuts none of the dominoes.

EXAMPLE 2.52. In Example 2.23 we proved that the grid dividing a rectangle into $m \times n$ squares 1×1 defines the total of $\binom{m+1}{2} \cdot \binom{n+1}{2}$ rectangles. If the original figure is a square, i.e., if $m = n$, then the total is $\binom{n+1}{2}^2$.

How many rectangles are defined by the upper left-hand $r \times r$ square of the $n \times n$ grid if we require the rectangles to touch at least one of the internal sides of the $r \times r$ square (see Fig. 2.7)?

SOLUTION: The rectangles in Example 2.52 belong to the upper left-hand $r \times r$ square, but not to the upper left-hand $(r-1) \times (r-1)$ square. Hence their number is

$$\binom{r+1}{2}^2 - \binom{r}{2}^2 = \frac{r^2}{4}\left((r+1)^2 - (r-1)^2\right) = r^3$$

NOTE: The total number of rectangles $\binom{n+1}{2}^2$ can be written as the sum of numbers for upper left-hand $r \times r$ squares for all values of r ($r = 1, 2, \ldots, n$). Thus we find that:

$$1^3 + 2^3 + \ldots + n^3 = \left(\frac{n(n+1)}{2}\right)^2$$

On the other hand since:

$$1 + 2 + \ldots + n = \frac{n(n+1)}{2}$$

we just found a combinatorial proof of an interesting identity:

$$(1 + 2 + \ldots + n)^2 = 1^3 + 2^3 + \ldots + n^3$$

EXAMPLE 2.53. How many ways can eight rooks be placed on a chess board so that they do not attack each other and none of them occupies a field on the main diagonal (the one connecting A-1 and H-8)?

RESULT: From the formula for derangements:

$$D_8 = \left\lfloor \frac{8!}{e} + \frac{1}{2} \right\rfloor = 14833$$

EXAMPLE 2.54. Prove that:

$$\binom{n}{0} + \binom{n}{2} + \binom{n}{4} + \ldots = \binom{n}{1} + \binom{n}{3} + \binom{n}{5} + \ldots = 2^{n-1}$$

COMBINATORIAL PROOF: Earlier we showed that:

$$\binom{n}{0} + \binom{n}{1} + \binom{n}{2} + \ldots + \binom{n}{n} = 2^n$$

In Example 1.39 we saw that the number of subsets of $A = \{a_1, \ldots, a_n\}$ with even cardinalities equals the number of its subsets with odd cardinalities.

The number of even subsets is

$$\binom{n}{0} + \binom{n}{2} + \binom{n}{4} + \ldots$$

The number of odd subsets is

$$\binom{n}{1} + \binom{n}{3} + \binom{n}{5} + \ldots$$

Since these add up to 2^n, there are $2^n/2 = 2^{n-1}$ of each kind.

ALGEBRAIC PROOF: In Newton's binomial formula, set $a = b = 1$ to find

$$\binom{n}{0} + \binom{n}{1} + \binom{n}{2} + \ldots + \binom{n}{n} = 2^n$$

while $a = 1$ and $b = -1$ yields

$$\binom{n}{0} - \binom{n}{1} + \binom{n}{2} - \ldots + (-1)^n \binom{n}{n} = 0$$

Adding and subtracting these two identities yields

$$\binom{n}{0} + \binom{n}{2} + \binom{n}{4} + \ldots = 2^{n-1}$$

and

$$\binom{n}{1} + \binom{n}{3} + \binom{n}{5} + \ldots = 2^{n-1}$$

EXAMPLE 2.55. Show that:

$$\binom{n}{k} = \binom{n-1}{k-1} + \binom{n-2}{k-1} + \ldots + \binom{k-1}{k-1}$$

COMBINATORIAL PROOF:

$$\binom{n}{k} = \text{number of } k\text{-element subsets among which the smallest is } 1$$
$$+\text{number of } k\text{-element subsets among which the smallest is } 2$$
$$\vdots$$
$$+\text{number of } k\text{-element subsets among which the smallest is } n-k+1$$
$$= \binom{n-1}{k-1} + \binom{n-2}{k-1} + \ldots + \binom{k-1}{k-1}$$

ALGEBRAIC PROOF: This identity is easily proved by mathematical induction. For the first several cases:

$$\binom{1}{1} = \binom{0}{0}$$
$$\binom{2}{1} = \binom{1}{0} + \binom{0}{0}$$
$$\binom{2}{2} = \binom{1}{1}$$

Assume

$$\binom{n}{k} = \binom{n-1}{k-1} + \binom{n-2}{k-1} + \ldots + \binom{k-1}{k-1}$$

Then using Pascal's formula

$$\binom{n+1}{k} = \binom{n}{k-1} + \binom{n}{k}$$
$$= \binom{n}{k-1} + \binom{n-1}{k-1} + \binom{n-2}{k-1} + \ldots + \binom{k-1}{k-1}$$

ANOTHER PROOF: The third solution is based on the generating function for the sequence of binomial coefficients:

$$\begin{aligned}(1+x)^n &= (1+x)(1+x)^{n-1} \\ &= (1+x)^{n-1} + x(1+x)^{n-1} \\ &= (1+x)^{n-2} + x(1+x)^{n-2} + x(1+x)^{n-1} \\ &\vdots \\ &= (1+x)^{k-1} + x(1+x)^{k-1} + \ldots + x(1+x)^{n-2} + x(1+x)^{n-1}\end{aligned}$$

Equating the coefficients of x^k we obtain

$$\binom{n}{k} = 0 + \binom{k-1}{k-1} + \ldots + \binom{n-2}{k-1} + \binom{n-1}{k-1}$$

EXAMPLE 2.56 (LEIBNIZ'S FORMULA). If $h^{(r)}(x)$ denotes the rth derivative of $h(x)$, show that the nth derivative of the product of functions $f(x)$ and $g(x)$ (having all necessary derivatives) can be expressed using Leibniz's formula:

$$(f \cdot g)^{(n)} = \binom{n}{0} f^{(n)} g^{(0)} + \binom{n}{1} f^{(n-1)} g^{(1)} + \ldots + \binom{n}{n} f^{(0)} g^{(n)}$$

HINT: Use mathematical induction.

EXAMPLE 2.57 (VANDERMONDE'S FORMULA). For an integer $r > 0$ the falling factorial of a real number a is

$$a^{\underline{r}} = a(a-1)\ldots(a-r+1)$$

In addition:

$$a^{\underline{0}} = 1$$

Prove Vandermonde's formula:

$$(x+y)^{\underline{n}} = \binom{n}{0} x^{\underline{n}} y^{\underline{0}} + \binom{n}{1} x^{\underline{n-1}} y^{\underline{1}} + \ldots + \binom{n}{n} x^{\underline{0}} y^{\underline{n}}$$

HINT: Use mathematical induction.

EXAMPLE 2.58 (FERMAT'S LESSER THEOREM). Assume a very large number of balls are available, each of which is colored by one of n colors. Also let p be some prime number. Find the number of different circular arrangements of p balls except those where all balls are the same color.

SOLUTION: If the balls are arranged in a line, there will be $n^p - n$ different arrangements. Since p is a prime, exactly p different linear arrangements correspond to one circular arrangement; therefore the solution is

$$\frac{n^p - n}{p}$$

NOTES: This is a combinatorial proof of Fermat's lesser theorem from the number theory, which states that:

For $n \in N$ and arbitrary prime p:

$$n^p \equiv n \pmod p$$

That is, the numbers n^p and n have the same remainder after division by p.

NOTES: Why is it important to consider only prime p? Does it follow from a similar problem that does not exclude arrangements of equally colored balls that n^p is divisible by p? Let us also mention that Fermat's lesser theorem is a special case of Euler's theorem, already mentioned in Example 2.30. See Chapter 3 for more about these theorems.

EXAMPLE 2.59. How many ways are there of selecting two disjoint subsets of the n-element set A?

SOLUTION: Let the first subset have k elements, $0 \leq k \leq n$. It can be chosen in $\binom{n}{k}$ ways. The second subset must be disjoint with the first; therefore its elements can be chosen from $(n-k)$ remaining elements of A. The total number of choices for the second set is then 2^{n-k}. Hence if the first subset has k elements, the total number of choices of two disjoint subsets of A is $\binom{n}{k} \cdot 2^{n-k}$. When we sum the corresponding numbers for all allowable ks, i.e., for $k = 0, 1, \ldots, n$ we have

$$\sum_{k=0}^{n} \binom{n}{k} 2^{n-k} = 3^n$$

The last equality was obtained from Newton's binomial formula, with $a = 2$, $b = 1$.

We are not done yet, however. Except in the case when both sets are empty, we counted every other choice twice. Therefore the final solution is

$$\frac{3^n + 1}{2} \qquad \square$$

The following three examples show three important probability distributions used in statistical physics. In all three examples we:

1. Start from simple physical models and assumptions.
2. Use combinatorics to find the numbers of different states of systems as functions of particular distributions of particles over the energies.
3. Find those distributions that maximize the numbers of possible states.

The distributions thus obtained are most often involved in macroscopic measurements. Equivalently these are distributions with maximum entropy.

EXAMPLE 2.60 (MAXWELL–BOLTZMANN STATISTICS). The following assumptions form this model:

- The particles are distinguishable.
- There are n_i particles with energy E_i, where $i = 1, 2, 3, \ldots$
- Every energy level E_i has g_i sublevels (phase cells).
- The number of particles that can be found at sublevels with energy E_i is limited by only the number of particles n_i on that level.
- The following conditions are satisfied

$$n_1 + n_2 + n_3 + \ldots = N \qquad n_1E_1 + n_2E_2 + n_3E_3 + \ldots = E_{tot}$$

According to these assumptions, if the distribution of particles over the energies is known, i.e., if the numbers n_i are known, the number of different arrangements of the particles is

$$P_{MB}(n_1, n_2, n_3, \ldots) = \frac{N!}{n_1!\,n_2!\,n_3!\ldots} g_1^{n_1} g_2^{n_2} g_3^{n_3} \ldots$$

To find the particular distribution that maximizes the number of states P_{MB}, we introduce two approximations. We justify these by the fact that the system typically contains very large numbers of particles, e.g., $N \approx 10^{23}$, and every level has many sublevels, e.g., $g_i \approx 10^8$.

For large values of r, Stirling's approximation for $r!$:

$$r! \sim \sqrt{2\pi r}\left(\frac{r}{e}\right)^r$$

yields

$$\ln r! \approx r \ln r - r$$

Later instead of a discrete distribution n_i over energies E_i $(i = 1,2,3,\dots)$, we use a continuous distribution $n(E)$.

Based on the preceding assumptions and using the method of Lagrange multipliers, we find that the number of states P_{MB} is maximized by:

$$n(E) = \frac{N}{kT}e^{-E/kT}$$

where N is the total number of particles in the system, k is Boltzmann's constant, and T is the absolute temperature of the system.

This distribution was first discovered by Boltzmann in 1896. Maxwell's distribution of particles over the velocities, first derived by Maxwell in 1860, can be derived from it.

EXAMPLE 2.61 (BOSE–EINSTEIN STATISTICS). This model is based on the following assumptions:

- The particles are indistinguishable.
- There are n_i particles with energy E_i, where $i = 1,2,3,\dots$
- Every energy level E_i has g_i sublevels.
- The number of particles that can be found at sublevels with energy E_i is limited by only the number of particles n_i on that level.
- The following conditions are satisfied

$$n_1 + n_2 + n_3 + \dots = N \qquad n_1E_1 + n_2E_2 + n_3E_3 + \dots = E_{tot}$$

From these assumptions we find

$$P_{BA}(n_1,n_2,n_3,\dots) = \frac{(n_1+g_1-1)!}{n_1!\,(g_1-1)!} \cdot \frac{(n_2+g_2-1)!}{n_2!\,(g_2-1)!} \cdot \frac{(n_3+g_3-1)!}{n_3!\,(g_3-1)!} \cdot \dots$$

This time the distribution $n(E)$ maximizing the number of states P_{BA} is

$$n(E) \sim \frac{1}{e^{(E-E_B)/kT} - 1}$$

where E_B is the Bose energy (Bose level) and T is the absolute temperature. Particles obeying this statistic are called bosons. For example photons and atoms of helium He^4 at low temperatures are bosons. At temperatures above several kelvins, the Bose–Einstein distribution for He^4 atoms becomes practically indistinguishable from the classical Maxwell–Boltzmann distribution.

This distribution was first found by Bose in 1924; it was published with help from Einstein.

EXAMPLE 2.62 (FERMI–DIRAC STATISTICS). This model is based on the following assumptions:

- The particles are indistinguishable.
- There are n_i particles with energy E_i, where $i = 1,2,3,\dots$
- Every energy level E_i has g_i sublevels.
- The number of particles that can be found at sublevels with energy E_i is limited by Pauli's exclusion principle: Every sublevel may contain at most one particle. (The assumption about the indistinguishability of particles is actually the second part of Pauli's principle.)
- The following conditions hold

$$n_1 + n_2 + n_3 + \dots = N \qquad n_1E_1 + n_2E_2 + n_3E_3 + \dots = E_{tot}$$

Based on the preceding we find

$$P_{FD}(n_1,n_2,n_3,\dots) = \binom{g_1}{n_1} \cdot \binom{g_2}{n_2} \cdot \binom{g_3}{n_3} \cdot \dots$$

The distribution that maximizes the number of states P_{FD} is

$$n(E) \sim \frac{1}{e^{(E-E_F)/kT} + 1}$$

where E_F is the so-called Fermi energy (Fermi level) and T is the absolute temperature. Particles obeying this distribution are called fermions. For example electron gas in metals obeys it even at room temperatures. Only at temperatures of 10^4 K does electron gas behave as an ideal gas, i.e., according to the Maxwell–Boltzmann distribution. Helium He^3 atoms at temperatures of a few kelvins are fermions, while at higher temperatures differences between Fermi–Dirac and Maxwell–Boltzmann distributions disappear.

This distribution was discovered by Fermi and Dirac in 1926.

EXAMPLE 2.63 (HAMMING'S FORMULA). Digital systems for handling information have many advantages over analog systems. One such advantage is the possibility of protection from damaging noise. A few instances where these error-correcting systems are used include CD players, computer and communication networks, and interplanetary image and data transfer.

We consider here a very simple code capable of error detection and even error correction.

Suppose we examine in detail information transfer conditions, then decide to use the transmitter of a power sufficient to ensure that when any of 2^n possible length-n binary words is transmitted, there is a small probability that some binary digits will be received incorrectly. Let us assume that the transmitter power is chosen so that the probability of more than r errors happening is practically zero.

If we choose to use only M length-n binary words out of 2^n, then if at the receiver we receive some words not in our vocabulary, we can be sure that an error occurred. This is called error detection. We can do even more, however. If we carefully pick the number M and words in the vocabulary, then we can also correct the occurrence of r (or less) errors. For example let $n = 8$, and let us find M such that our code can correct $r = 2$ errors. If one of the words is 00111100, then our system will operate so that if it receives some word from the 2-neighborhood of 00111100, these all being words that differ from this one at 0, 1, or 2 binary places, it recognizes that the originally transmitted word was 00111100.

Every r-neighborhood has the following number of words:

$$\binom{n}{0} + \binom{n}{1} + \ldots + \binom{n}{r}$$

Hence M must satisfy the Hamming formula:

$$M \leq \frac{2^n}{\binom{n}{0} + \binom{n}{1} + \ldots + \binom{n}{r}}$$

For the system in our example to correct $r = 2$ errors, among all $2^8 = 256$ possible words, we can use at most $M = \lfloor 256/(1+8+28) \rfloor = 6$ binary words.

The problem of choosing those M words cannot be discussed here. Let us just note that the number of binary digits that differ in two binary words is called the Hamming distance of these two words.

EXAMPLE 2.64 (TOWERS OF HANOI). One of three vertical sticks has n disks, so placed that the largest disk is at the bottom and remaining disks are stacked in decreasing order. The problem is to move the disks to another stick in as few moves as possible and with an additional requirement: At all times, each stick must have disks stacked in decreasing order.

The story following this 1883 problem by Lucas relates that at the beginning of time, God Brahma put 64 golden disks on one of three diamond needles, then ordered his priests to move them to another needle and always to be careful not to put larger disk above smaller. When the priests finish their job, Brahma's tower will crumble, and the world will end.

SOLUTION: Let T_n denote the minimum number of moves needed to move n disks according to the preceding rules. According to these rules, at the moment we move the largest disk from Stick 1 to Stick 3, all other disks must be stacked at Stick 2 in decreasing order. To accomplish this, we needed T_{n-1} moves. These must be followed by moving the largest disk and by additional T_{n-1} moves to place $(n-1)$ smaller disks over it (see Fig. 2.8). Thus we obtain the following recursion:

$$T_n = 2T_{n-1} + 1 \quad (n > 1) \qquad T_1 = 1$$

The solution of this recursion is $T_n = 2^n - 1$, which means that Brahma's priests can accomplish their task in $2^{64} - 1$ moves. If each move requires 1 second, the time needed is around $5.85 \cdot 10^{11}$ years!

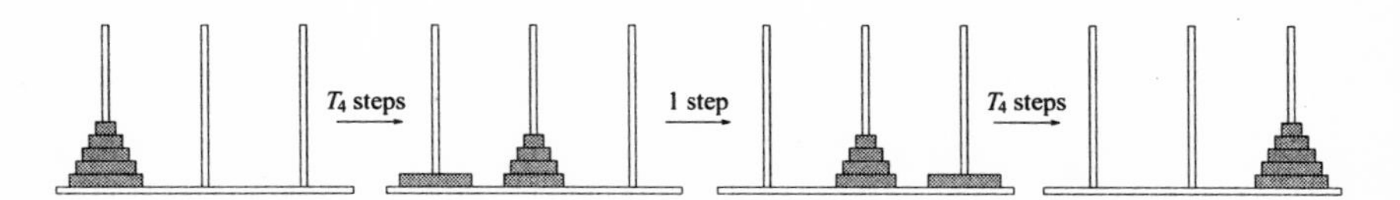

FIGURE 2.8. Towers of Hanoi.

Let us look at several ways of solving this recursion. In this particular example, only one method is the most elegant; however we consider other methods as well.

APPROACH 1: Apply the recursion several times, until we notice some regularity:

$$\begin{aligned}T_n &= 2T_{n-1}+1\\ &= 4T_{n-2}+2+1\\ &= 8T_{n-3}+4+2+1\\ &\vdots\\ &= 2^{n-1}T_1+2^{n-2}+\ldots+4+2+1\end{aligned}$$

Then recall $T_1 = 1$ and $2^{n-2}+\ldots+4+2+1 = 2^{n-1}-1$ to finally write $T_n = 2^n - 1$.

APPROACH 2: The recursion $T_n = 2T_{n-1}+1$ can be written as:

$$T_n + 1 = 2(T_{n-1}+1)$$

Therefore with $U_n = T_n + 1$, we obtain an auxiliary recursion $U_n = 2U_{n-1}$, $U_1 = 2$, which is easy to solve: $U_n = 2U_{n-1} = 4U_{n-2} = \ldots = 2^{n-1}U_1 = 2^n$.. Finally

$$T_n = U_n - 1 = 2^n - 1$$

APPROACH 3: Recursion $T_n = 2T_{n-1}+1$ and the initial condition $T_1 = 1$ can be used to form the generating function of $\{T_n\}$ (the recursion suggests $T_0 = 0$):

$$\begin{aligned}T(x) &= T_0 + T_1x + T_2x^2 + T_3x^3 + \ldots + T_nx^n + \ldots\\ 2xT(x) &= 2T_0x + 2T_1x^2 + 2T_2x^3 + \ldots + 2T_{n-1}x^n + \ldots\end{aligned}$$

Hence:

$$T(x) - 2xT(x) = T_0 + x + x^2 + x^3 + \ldots + x^n + \ldots$$

That is,

$$T(x) = \frac{x}{(1-2x)(1-x)}$$

Since

$$\frac{x}{(1-2x)(1-x)} = \frac{1}{1-2x} - \frac{1}{1-x}$$

we find

$$T(x) = (1 + 2x + (2x)^2 + (2x)^3 + \ldots + (2x)^n + \ldots) \\ -(1 + x + x^2 + x^3 + \ldots + x^n + \ldots)$$

Finally:

$$T(x) = (2-1)x + (2^2-1)x^2 + (2^3-1)x^3 + \ldots + (2^n-1)x^n + \ldots$$

which implies

$$T_n = 2^n - 1$$

EXAMPLE 2.65 (POLYNOMIAL FORMULA). This example shows the combinatorial derivation of the formula for the power of a polynomial:

$$(a_1 + a_2 + \ldots + a_r)^n$$

A typical term in its full expansion has the following form:

$$a_1^{k_1} a_2^{k_2} \ldots a_r^{k_r}$$

where $k_1 + k_2 + \ldots + k_r = n$. The only remaining problem is to find how many times each of these terms appears.

Term a_1 can be selected k_1 times from n boxes, i.e., in $\binom{n}{k_1}$ ways. Then term a_2 can be selected k_2 times from the remaining $n - k_1$ boxes, i.e., in $\binom{n-k_1}{k_2}$

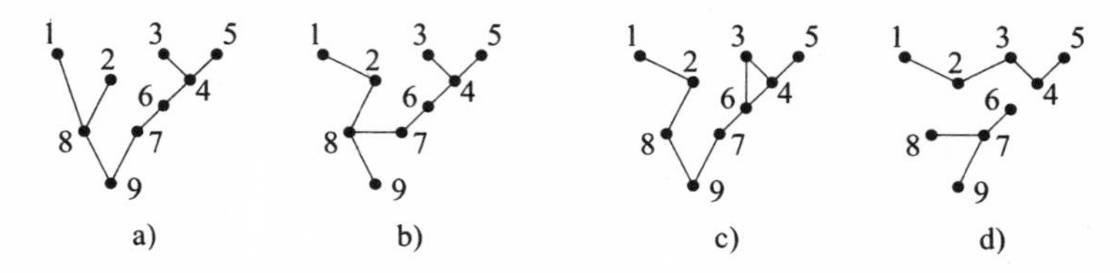

FIGURE 2.9. Trees and nontrees. Graphs a and b are trees. Graph c is not, because of the 3–4–6–3 contour. Graph d consists of two trees, but it is not a tree because it is not connected.

ways. Continuing this process, we find that the term $a_1^{k_1}a_2^{k_2}\dots a_r^{k_r}$ appears the following number of times:

$$\binom{n}{k_1}\binom{n-k_1}{k_2}\cdots\binom{n-k_1-\ldots-k_{r-1}}{k_r} = \frac{n!}{k_1!k_2!\ldots k_r!}$$

If, analogous to the binomial formula and the binomial coefficients, we define

$$\binom{n}{k_1,k_2,\ldots,k_r} = \frac{n!}{k_1!k_2!\ldots k_r!} \qquad \begin{matrix} k_1,k_2,\ldots,k_r \geq 0 \\ k_1+k_2+\ldots+k_r = n \end{matrix}$$

then we can write the polynomial formula

$$(a_1+a_2+\ldots+a_r)^n = \sum_{\substack{k_1,k_2,\ldots,k_r\geq 0 \\ k_1+k_2+\ldots+k_r=n}} \binom{n}{k_1,k_2,\ldots,k_r} a_1^{k_1}a_2^{k_2}\ldots a_r^{k_r}$$

which was first discovered by Leibniz.

EXAMPLE 2.66 (CAYLEY'S THEOREM). Trees are special connected graphs, distinguished by the fact that no edges of the tree form a closed contour. Figure 2.9 shows two graphs that are trees, and two that are not.

It is easily shown that trees with n vertices have $n-1$ edges. A more difficult question however is how many trees can be formed over the set of vertices $\{v_1,v_2,\ldots,v_n\}$?

First let us find how many trees are such that d_i edges are connected at the vertex v_i $(i=1,2,\ldots,n)$. (We also say that the degree of vertex v_i is d_i.)

Note that when we add all vertex degrees, each edge is counted exactly twice, therefore

$$d_1+d_2+\ldots+d_n = 2(n-1)$$

The next step is the inductive proof (not quite trivial) that there are

$$\binom{n-2}{(d_1-1),\ldots,(d_n-1)}$$

trees whose vertex v_i has degree d_i $(i=1,\ldots,n)$, so that the total number of trees with n vertices is

$$\sum_{\substack{d_1,d_2,\ldots,d_n\geq 1\\ d_1+d_2+\ldots+d_n=2(n-1)}} \binom{n-2}{(d_1-1),\ldots,(d_n-1)}$$

This sum can be simplified if we write

$$k_i = d_i - 1 \qquad (i=1,2,\ldots,n)$$

then use the polynomial formula for $a_1 = \ldots = a_n = 1$:

$$\sum_{\substack{k_1,k_2,\ldots,k_n\geq 0\\ k_1+k_2+\ldots+k_n=n-2}} \binom{n-2}{k_1,\ldots,k_n} = (1+\ldots+1)^{n-2} = n^{n-2}$$

This ends the proof of Cayley's theorem.

NOTE: Enumerating all isomers of the saturated carbo-hydrates C_nH_{2n+2} is a much more difficult problem, and it requires using Polya's theorem because while Cayley's theorem holds for graphs whose vertices are distinguishable (labeled graph), carbon and hydrogen atoms are not distinguishable inside their species (unlabeled graph).

EXAMPLE 2.67. 1. How many elements does $P(A)$, the partitive set of the n-element set A, have? How many elements are there in the Cartesian product $A \times A$? How many relations are there over the set A?

2. How many reflexive relations exist over an n-element set?

3. How many symmetric relations exist over an n-element set?

4. How many antisymmetric relations exist over an n-element set?

5. How many relations that are both symmetric and antisymmetric exist over an n-element set?

RESULTS:

1. $2^n, n^2, 2^{n^2} = 2^n \cdot 4^{\binom{n}{2}}$.

2. $2^{n^2-n} = 4^{\binom{n}{2}}$.

3. $2^n \cdot 2^{\binom{n}{2}}$.

4. $2^n \cdot 3^{\binom{n}{2}}$.

5. 2^n.

EXAMPLE 2.68. Let $P(m,n)$ be the number of ways of obtaining m as the sum after rolling n differently colored dice. Determine the generating function for the sequence $\{P(m,n)\}$.

SOLUTION: Since the dice are colored differently the following outcomes are considered as different

$$1,1,\ldots,1,2 \qquad 2,1,\ldots,1,1$$

With a little inspiration, we find the generating function:

$$G(x) = (x+x^2+x^3+x^4+x^5+x^6)^n$$

where the coefficient next to x^m represents the number of outcomes having the sum m.

EXAMPLE 2.69. Assume we are rolling n mutually indistinguishable dice at once.

1. How many different outcomes are possible?

2. How many ways are there of obtaining the outcome described by the multiset $\langle 1,1,\ldots,1,2\rangle$?

3. How many different sums are possible?

4. How many different outcomes give sum m? Which sum is produced by the largest number of outcomes? Is that sum the most probable?

SOLUTIONS:

1. Since the dice are indistinguishable and each can have an outcome $1,2,3,4,5$, or 6, we are working with the n-combinations with repetitions of a six-element set, and the answer is $\binom{n+5}{5}$.

2. This outcome is much more probable than the outcome where all dice show 1, because all 1s can be obtained in only one way. The outcome described by the multiset $\langle 1,1,\ldots,1,2\rangle$ can be obtained in n ways, although we cannot distinguish among them. (The dice are indistinguishable, and they are rolled simultaneously).

3. All sums from $1+1+\ldots+1=n$ to $6+6+\ldots+6=6n$ are possible, so the answer is $6n-n+1=5n+1$.

4. If $m<n$, or $m>6n$, none of the outcomes can have the sum m. Only one outcome yields sum $m=n$; also only one yields the sum $m=n+1$, although the latter is much more probable. Two outcomes produced the sum $m=n+2$. They are described by multisets $\langle 1,1,\ldots,1,3\rangle$ and $\langle 1,1,\ldots,2,2\rangle$. Thus while the sequence begins as the Fibonacci sequence:

$$0,\ldots,0,1,1,2,3,5$$

that is where similarities stop, because the next numbers are 7, 10, 13, 18, 23, 29, 35, ...

With a little luck and imagination we come up with the following generating function:

$$G(x)=(1+x+x^2+\ldots)(1+x^2+x^4+\ldots)\ldots(1+x^6+x^{12}+\ldots)$$

We see that next to x^m it gives the number of all possible partitions of m using an arbitrary number of terms $1,2,3,4,5$, or 6. How do we extract the number of partitions using exactly n terms $1,2,3,4,5$, or 6 from this number? The solution is to introduce an additional variable:

$$G(x,y)=(1+xy+x^2y^2+\ldots)(1+x^2y+x^4y^2+\ldots)\ldots$$
$$\ldots(1+x^6y+x^{12}y^2+\ldots)$$

Now the number of outcomes producing a sum m after rolling n identical dice is the coefficient next to $x^m y^n$ in the expansion:

$$G(x,y) = \frac{1}{1-xy}\frac{1}{1-x^2y}\cdots\frac{1}{1-x^6y}$$

Using the Taylor expansion in two variables, we can write

$$B_{m,n} = \frac{1}{m!n!}\left(\frac{\partial^m}{\partial x^m}\frac{\partial^n}{\partial y^n}G(x,y)\right)_{x=y=0}$$

However this formula is not very useful for determining numbers $B_{m,n}$. Much more useful is the observation that just as polynomial multiplication is easily performed as the convolution of sequences of polynomial coefficients, multiplication of polynomials with two variables can be accomplished as the convolution of matrices of corresponding coefficients. These calculations are easily performed on a computer.

3

Number Theory

Since antiquity, people have been interested in the properties of numbers. For example in Babylon, a whole millennium before Pythagoras, mathematicians knew how to determine *Pythagorean numbers*, i.e., the integers that represent the sides of a right-angled triangle. Theory of numbers answers many other questions as well, yet there are many unsolved problems. The publicity given to the successful collective attack by thousands of Internet users on the Rivest–Shamir–Adleman (RSA) public coding system in spring 1994 shows how important a part of mathematics number theory is.

3.1. Divisibility of Numbers

In this section we consider the divisibility of numbers, the division remainders, and similar elementary number theory concepts.

Basic Notions and Theorems

DEFINITION 3.1 (DIVISIBILITY). Integer a is divisible by integer $b \neq 0$ if there exists integer q, such that $a = bq$.

If a is divisible by b, we write $b \mid a$ (read: *b divides a*).
Definition 3.1 is written symbolically as:

$$(\forall a, b \in Z,\ b \neq 0) \quad b \mid a \Leftrightarrow (\exists q \in Z) \quad a = bq$$

If a is not divisible by b, we write $b \nmid a$ (*b does not divide a*), for example:

$$3 \mid 15 \qquad 6 \nmid 15$$

We can prove the following properties of integers.

THEOREM 3.1. *Let a, b, and c be any integers. Then:*

a. $b \mid a \Rightarrow (\forall m \in Z)\ b \mid am.$

b. $a \mid b \wedge b \mid c \Rightarrow a \mid c.$

c. $a \mid b \wedge a \mid c \Rightarrow (\forall m, n \in Z)\ a \mid (mb + nc).$

PROOF:

a. $b \mid a \Rightarrow (\exists q \in Z)\ a = bq$. Therefore:

$$(\forall m \in Z)\ am = bqm = bq_1 \Rightarrow b \mid am$$

b. $a \mid b \wedge b \mid c \Rightarrow b = q_1 a \wedge c = q_2 b \Rightarrow c = q_1 q_2 a = qa \Rightarrow a \mid c$.

c. $a \mid b \wedge a \mid c \Rightarrow b = q_1 a \wedge c = q_2 a$. Therefore $(\forall m, n \in Z)$:

$$mb + nc = mq_1 a + nq_2 a = (mq_1 + nq_2)a = qa \Rightarrow a \mid (mb + nc). \square$$

In the following we learn that if we take two integers a and b $(b \neq 0)$, and divide a/b, then there is only one integer q (*quotient*) and only one nonnegative integer r less than $|b|$ (*remainder*), such that:

$$a = bq + r \qquad 0 \leq r < |b| \tag{3.1}$$

We see very soon why the uniqueness of numbers q and r is very important. For the sake of simplicity, we prove the following theorem for the case when $b > 0$. The proof for $b < 0$ is completely analogous.

THEOREM 3.2 (DIVISION). *If $a \in Z$ and $b \in N$, then a can be uniquely represented as:*

$$a = bq + r \qquad (q, r \in Z \quad 0 \leq r < b)$$

PROOF: First we must prove the existence of q and r. After that we prove their uniqueness:

- Existence: Let us find *the least nonnegative* number among the following numbers $\ldots, a - 2b, a - b, a, a + b, a + 2b, \ldots$ Let it be $r = a - qb$ for some $q \in Z$. Then:

  $$a = bq + r$$

  Since r was picked as the *least nonnegative* among the numbers $\ldots, a - 2b, a - b, a, a + b, a + 2b, \ldots$, we see that $a - (q + 1)b < 0$. This means that $0 \leq r < b$. This proves the existence of numbers q and r.

- Uniqueness: Suppose numbers q' and r' also satisfy the conditions of the theorem, i.e.:

$$a = bq' + r' \qquad (0 \le r' < b)$$

If we subtract this equation from the equation $a = bq + r$, we obtain

$$0 = b(q - q') + (r - r')$$

This implies that $b(q - q') = r' - r$, i.e., $b \mid (r' - r)$. Since r and r' are nonnegative and less than b, their difference is also less than b. The only remaining possibility when b divides the difference $(r' - r)$ is when $r' - r = 0$, i.e., $r' = r$. Then $q' = q$, which proves the uniqueness of q and r. □

EXAMPLE 3.1. Later we focus considerable attention on properties shared by numbers having the same remainder when divided by the same number. For example let x be any number with r_1 as a remainder after division by m and let y be any number with r_2 as a remainder after division by m.

No matter which numbers x and y were chosen from the *residue classes* r_1 and r_2, the sum $z = x + y$ always has the same remainder after division by m. This remainder depends on only the initially chosen classes r_1 and r_2. A similar statement holds for the product $w = xy$.

Why is it so? From:

$$x = q_1 m + r_1 \qquad y = q_2 m + r_2$$

we find that:

$$z = x + y = (q_1 + q_2)m + (r_1 + r_2)$$

We say that z belongs to the same residue class as the sum $(r_1 + r_2)$. For the product $w = xy$, we have

$$w = xy = (q_1 m + r_1)(q_2 m + r_2) = (q_1 q_2 m + q_1 r_2 + q_2 r_1)m + r_1 r_2$$

Therefore w belongs to the same residue class as the product $r_1 r_2$.

EXAMPLE 3.2. Consider a decimal fraction $1/7 = 0.142857142857\ldots$ Why does its repetition period equal 6? Could it by any chance be greater?

SOLUTION: Division yields

```
1:7 = 0.1428571...
10
 30
  20
   60
    40
     50
      10
       ⋮
```

Had remainder 1 appeared earlier, the period would have been shorter. This way the period is 6. The period could not be greater than 6 because we already had all remainders other than 0 once. If some remainder other than 1 appeared for the second time, it would make the period smaller, not greater. If 0 appeared, the period would have been 1 because all other digits would have been zeros as well.

EXAMPLE 3.3. Lagrange showed that the sequence formed by the last digits of the Fibonacci numbers is periodic; the period is 60.

PROOF: In Example 3.2, repetition starts as soon as some remainder is repeated. In this example we consider the last digits of Fibonacci numbers, so repetition starts as soon as we find two ones, just as at the beginning of the sequence. This is so because the sequence is formed almost exactly like Fibonacci's sequence: The next number in the sequence is the sum of the previous two, but we register only the last digits of the numbers produced in that way (this kind of addition is called *addition modulo* 10): $1, 1, 2, 3, 5, 8, 13 \mapsto 3, 11 \mapsto 1, 4, 5, 9, 14 \mapsto 4, \ldots$ The sequence formed in this way is

$\underline{\mathbf{1}}, \underline{\mathbf{1}}$, 2, 3, 5, 8, 3, 1, 4, 5, 9, 4, 3, 7, 0, 7, 7, 4, 1, 5, 6, 1, 7, 8, 5, 3, 8, 1, 9, 0, 9, 9, 8, 7, 5, 2, 7, 9, 6, 5, 1, 6, 7, 3, 0, 3, 3, 6, 9, 5, 4, 9, 3, 2, 5, 7, 2, 9, 1, 0, $\underline{\mathbf{1}}, \underline{\mathbf{1}}, \ldots$ □

Primes and Canonical Decomposition

DEFINITION 3.2 (PRIME NUMBERS). A prime is an integer greater than 1 divisible only by itself and one. All other integers greater than 1 are called composite integers.

NOTE: Number 1 is neither a prime nor a composite. This is due to its special status as a neutral element (unit) in multiplication.

The sequence of primes begins as: 2, 3, 5, 7, 11, 13, 17, 19, ... The easiest way of continuing this sequence involves the ancient method known as the *sieve of Eratosthenes*: To determine all primes $< n$, first list all natural numbers from 2 to $n-1$. Then eliminate all even numbers except 2, then all numbers divisible by 3 except 3, and so on for 5, 7, 11, 13, etc., until we have covered all prime numbers $< \sqrt{n}$. Note: After eliminating even numbers, we need not worry about numbers divisible by 4, 6, 8, etc., because these are even numbers. Similarly after eliminating all numbers divisible by 3, we need not worry about those divisible by 6, 9, 12, etc., because they have already been eliminated. Thus to find all primes $< n$, we eliminate numbers divisible by primes $< \sqrt{n}$.

EXAMPLE 3.4 (PRIMES LESS THAN 100). To determine the primes less than 100, we use the sieve of Eratosthenes to eliminate all numbers divisible by 2, 3, 5, and 7. This procedure yields the following 25 numbers:

$$2, 3, 5, 7, 11, 13, 17, 19, 23, 29, 31, 37, 41, 43, 47, 53, 59,$$
$$61, 67, 71, 73, 79, 83, 89, 97$$

The answer to the question of how many primes exist was already known to Greek mathematicians. In the ninth book of the *Elements*, Euclid gives the following proof that the set of primes is infinite.

Suppose there are only finitely many primes. Denote these as $p_1, p_2, \ldots, p_n$. Let p_n be the greatest among these. If we form the number:

$$P = p_1 p_2 \cdots p_n + 1$$

we see that $P > p_n$, so there are only two possibilities for P:

- P is a prime, which contradicts our assumption that p_n is the greatest prime.

- P is a composite, which contradicts the assumption that $p_1, p_2, \ldots, p_n$ are the only primes, because when divided by any of them, P has remainder 1. Thus since P is a composite number, there must exist other primes that divide it.

This proves that there cannot be only finitely many primes:

THEOREM 3.3 (EUCLID'S THEOREM). *The set of primes is infinite.*

EXAMPLE 3.5 (ANOTHER PROOF). Euclid's theorem can be proved in many different ways. Here we add only the proof given by Kummer in 1878:

Let $p_1, p_2, \ldots, p_n$ be the only primes and let $N = p_1 p_2 \cdots p_n > 2$. The number $N-1$ has at least one prime factor, e.g., p_k, for some $1 \leq k \leq n$. Since $p_k \mid N$ and $p_k \mid (N-1)$, we see that p_k divides the difference $N-(N-1)=1$, which is impossible.

EXAMPLE 3.6 (EULER'S TRINOMIAL). In 1772 Euler gave the following quadratic trinomial:

$$x^2 + x + 41$$

which for $x = 0, 1, 2, \ldots, 39$ produces different primes.

NOTE: A useful formula for generating all primes does not exist. Some formulas are true if and only if their argument is prime; other formulas theoretically generate all primes, but none of these can be used because they either involve extremely complicated calculations or constants whose values depend on knowing all primes in advance.

EXAMPLE 3.7 (FERMAT NUMBERS). If $2^m + 1$ is a prime, then m must be a power of 2. Prove.

SOLUTION: Assume that m is not a power of 2 or in other words, that $m = 2^n \cdot k$ for some $n \geq 0$ and odd k. Then:

$$2^m + 1 = 2^{2^n \cdot k} + 1 = (2^{2^n})^k + 1$$

Since for odd k we have $a^k + 1 = (a+1)(a^{k-1} - a^{k-2} + a^{k-3} - \ldots + 1)$, we see that $(2^{2^n} + 1) \mid [(2^{2^n})^k + 1]$. Therefore if $2^m + 1$ is a prime, then $k = 1$; i.e., m is a power of 2.

NOTES: The numbers $F_n = 2^{2^n} + 1$ are called the Fermat numbers. In 1640 Fermat conjectured that all of these numbers were prime, probably his only incorrect conjecture. The first counterexample was identified by Euler, who found that $F_5 = 4294967297 = 641 \cdot 6700417$. Since that time, a lot of effort and in the last few decades, a lot of computer time has been dedicated to factoring Fermat numbers. Although Fermat's conjecture is known to be false, it is not known whether there are any primes among the Fermat numbers except F_0, F_1, F_2, F_3, and F_4.

One of the many properties of Fermat numbers is the following formula: $F_n = F_0F_1 \dots F_{n-1} + 2$, which implies that the only common factor that they might have is 2. But since Fermat numbers are odd, this is not the case, therefore they are all relatively prime. This observation is the basis of Polya's proof of the infiniteness of the set of primes.

Fermat numbers emerge in quite an unexpected place — Gauss's theorem about the constructibility of regular polygons*: *A regular polygon with n sides can be constructed using a ruler and compass if and only if* $n = 2^r p_1 \dots p_m$, *where* $r \geq 0$, *and* $p_1, \dots, p_m$ *are different Fermat primes or* $n = 2^r$, *where* $r \geq 2$. Among other things, this theorem shows that it is impossible to construct a regular heptagon ($n = 7$), but constructing a regular heptadecagon ($n = 17$) is possible. The question of constructing regular polygons with a ruler and compass was considered even by Ancient Greek mathematicians, who constructed the regular pentagon but stopped there.

EXAMPLE 3.8. Prove that all Fermat numbers of order higher than 1 have 7 as the last digit.

SOLUTION: All numbers of the form 2^{2^n}, where $n > 1$, end in the digit 6 because after $2^{2^2} = 16$, all others are produced by squaring the previous number of that form. This is the key observation here, because squaring a number ending in 6 again yields a number ending in 6:

$$(10k+6)^2 = 10(10k^2 + 12k) + 6$$

EXAMPLE 3.9. If $p > 3$ is a prime, then $p^2 - 1$ is divisible by 24. Prove.

SOLUTION: Since p is an odd number, $(p-1)$ and $(p+1)$ are consecutive even numbers. Thus one of these is divisible by 2 and the other by 4. Therefore $p^2 - 1 = (p-1)(p+1)$ is a number divisible by 8. Similarly since $(p-1), p,$

*Gauss made this discovery when he was only nineteen.

and $(p+1)$ are consecutive numbers, one is divisible by 3, but it cannot be p, because it is a prime greater than 3. This ends the proof.

We may think that the following theorem (Theorem 3.4) does not require proof because our experience guarantees its correctness, but we must be careful, since in mathematics experience is not a proof. Theorem 3.4 is so important that it is often called the *Fundamental Theorem of Arithmetic*.

THEOREM 3.4 (FUNDAMENTAL THEOREM OF ARITHMETIC). *Every integer greater than* 1 *has a unique representation as a product of prime factors. (Their order is considered to be irrelevant.)*

PROOF: First we prove the *existence* of such a representation. By definition all primes are represented in this way, and from experience we know that the first several composites are similarly represented: $4 = 2^2$, $6 = 2 \cdot 3$, $8 = 2^3$, $10 = 2 \cdot 5, \ldots$ If we assume there are composites without such representation, there must exist the smallest among them. Let us denote it by m. Since m is a composite, we can write $m = m_1 m_2$, where $1 < m_1, m_2 < m$. Since m is assumed to be the least composite without factorization into primes, m_1 and m_2 certainly have unique representation as a product of prime factors, but then m, too, has it, which is a contradiction.

Now we prove the *uniqueness* of prime factorization. For all primes and the first several composites, we know it is unique. Now suppose there are composites with two or more prime factorizations. Denote the least such composite by n. Then we can write

$$n = p_1 p_2 \ldots p_r \qquad n = q_1 q_2 \ldots q_s$$

where $p_1 \leq p_2 \leq \ldots \leq p_r$ and $q_1 \leq q_2 \leq \ldots \leq q_s$. It is easy to see that $p_i \neq q_j$, because if these two factorizations had a common factor p, after cancellation by p we would have two different factorizations for n/p, an integer less than n. Hence we can take $p_1 < q_1$.

Consider now $k = p_1 q_2 \ldots q_s$. We now show that if n has two different factorizations, then $n - k$, too, has a nonunique factorization. This again contradicts the assumed minimality of n.

Indeed since both n and k are divisible by p_1, so is $n - k$; hence there exists a factorization $n - k = p_1 f_2 \ldots f_a$. On the other hand, $n - k = (q_1 - p_1) q_2 \ldots q_s = g_1 \ldots g_b q_2 \ldots q_s$. Since $p_1 \neq q_j$, these two factorizations differ because the former includes p_1, but the latter does not. This proves the uniqueness of the so-called *canonical* factorization. □

NOTES: This important theorem is often proved through a sequence of lemmas and theorems that follow from the Euclidean algorithm (Theorems 3.8 and 3.18 and Example 3.10). Such a proof was used in the ninth book of the *Elements* of Euclid.

Even with the computing power and knowledge about integers we have today, factoring large numbers is still a formidable problem. Since number theory is used in information coding, factoring integers became a strategically important problem, too.

To illustrate that experience is not a sufficient proof, we mention that in the system of numbers* of the form $a+b\sqrt{-5}$, number 6 has two factorizations: $6=2\cdot 3=(1-\sqrt{-5})(1+\sqrt{-5})$. Also 2, 3, and $1\pm\sqrt{-5}$ cannot be factored any further.

EXAMPLE 3.10. An immediate consequence of the fundamental theorem of arithmetic is the following fact, which is often used in solving problems and proofs:

If a and m have no common factors and $m \mid ax$, then $m \mid x$.

In the traditional proof of the fundamental theorem of arithmetic, this result is actually a very important step in the sequence of lemmas and theorems that lead to the final proof. In that context it is proved using the Euclidean algorithm, and Theorem 3.18 in particular.

EXAMPLE 3.11 (NUMBER OF DIVISORS). The number of divisors of n is denoted by $\tau(n)$ or sometimes by $d(n)$. If the canonical representation of n is

$$n=p_1^{\alpha_1}p_2^{\alpha_2}\cdots p_k^{\alpha_k}$$

then every divisor of n has the following form:

$$d=p_1^{\beta_1}p_2^{\beta_2}\cdots p_k^{\beta_k}$$

*In 1952 Heegner showed that in a system of numbers of the form $a+b\sqrt{-D}$ factorization is unique only when $-D$ is one of the *Heegner* numbers: $-1,-2,-3,-7,-11,-19,-43,-67,-163$. Heegner numbers have many other interesting properties; e.g., for $a=e^{\pi\sqrt{D}}$ the number $a-744+196884/a-21493760/a^2+\ldots$ is a cube of an integer. For example, $e^{\pi\sqrt{163}}=640320^3+744-\epsilon$, where $\epsilon<10^{-12}$. Indeed $e^{\pi\sqrt{163}}=262537412640768743.99999999999925007\ldots$

where β_i $(i = 1,2,\ldots,k)$ are such that:

$$0 \le \beta_i \le \alpha_i \qquad (i = 1,2,\ldots,k)$$

Because of the uniqueness of the canonical representation, which applies to d as well, every divisor of n is uniquely determined by the choice of the exponents β_i $(i = 1,2,\ldots,k)$. Since β_i can be selected from $(\alpha_i + 1)$ different values, the total number of choices for d, i.e., the total number of divisors of n, is

$$\tau(n) = (\alpha_1 + 1)(\alpha_2 + 1)\ldots(\alpha_k + 1)$$

GCD, LCM, and Euclidean Algorithm

DEFINITION 3.3 (COMMON DIVISOR). The integer d for which $d \mid a$ and $d \mid b$ is a common divisor of integers a and b.

Since a divisor of a number cannot be greater than the number itself, we see that among the common divisors of a and b there exists the greatest.

DEFINITION 3.4 (GREATEST COMMON DIVISOR — GCD). The greatest element in the set of common divisors of integers a and b, where $a \neq 0$ or $b \neq 0$, is called the greatest common divisor of a and b, denoted $\text{GCD}(a,b)$ or simply (a,b).

If a and b are relatively prime, i.e., if they do not have common factors, their GCD is 1, i.e., $(a,b) = 1$. For that reason to emphasize that a and b are relatively prime, we write $(a,b) = 1$.

EXAMPLE 3.12. For example the important result mentioned in Example 3.10,

If a and m have no common factors and $m \mid ax$, then $m \mid x$,

is written symbolically as:

$$(a,m) = 1 \wedge m \mid ax \Rightarrow m \mid x$$

DEFINITION 3.5 (COMMON MULTIPLE). The integer s for which $a \mid s$ and $b \mid s$ is called a common multiple of integers a and b.

A multiple of a number cannot be smaller than the number itself; therefore there exists the least among the common multiples of integers a and b.

DEFINITION 3.6 (LEAST COMMON MULTIPLE — LCM). The least element in the set of common multiples of integers a and b is called the least common multiple, denoted $\text{LCM}[a,b]$ or simply $[a,b]$.

If a and b are relatively prime, then their LCM equals their product, i.e.:

$$(a,b) = 1 \Rightarrow [a,b] = ab$$

In fact because of the fundamental theorem of arithmetic, the more general Theorem 3.5 is true.

THEOREM 3.5 (GCD AND LCM). *If:*

$$a = p_1^{\alpha_1} p_2^{\alpha_2} \dots p_r^{\alpha_r} \qquad b = p_1^{\beta_1} p_2^{\beta_2} \dots p_r^{\beta_r}$$

where among the exponents α_i and β_i there may be some zeros, then:

$$(a,b) = GCD(a,b) = p_1^{\min(\alpha_1,\beta_1)} p_2^{\min(\alpha_2,\beta_2)} \dots p_r^{\min(\alpha_r,\beta_r)}$$
$$[a,b] = LCM[a,b] = p_1^{\max(\alpha_1,\beta_1)} p_2^{\max(\alpha_2,\beta_2)} \dots p_r^{\max(\alpha_r,\beta_r)}$$

The consequence of Theorem 3.5 is Theorem 3.6.

THEOREM 3.6. *The product of the* GCD *and* LCM *of two integers equals the absolute value of the product of those two numbers, i.e.:*

$$(a,b) \cdot [a,b] = |ab|$$

NOTE: Since factoring integers is a very difficult task, except when a and b are relatively small or easy to factor, we never use the results of Theorem 3.5 to determine their GCD. Although the formula is simple, the calculations involved are more complicated than the *Euclidean algorithm*, probably the oldest nontrivial algorithm still in use. The Euclidean algorithm first appeared in Theorems VII-1 and VII-2 of Euclid's *Elements*.

To prove and fully understand this algorithm we must examine the properties of GCD in an entirely different manner.

The key factor in the proof is Theorem 3.7.

THEOREM 3.7. *If $a = bq + r$, then:*

$$(a,b) = (b,r)$$

PROOF: The proof has two parts. If $d = (a,b)$ and $d' = (b,r)$, we first prove that $d \mid d'$, then that $d' \mid d$, which when taken together imply $d = d'$.

- If $d = (a,b)$, then $d \mid a$ and $d \mid b$, therefore $d \mid r = a - bq$. Thus, $d \mid b$, and $d \mid r$, which implies $d \mid d' = (b,r)$.
- Using similar arguments, we find that $d' \mid d$.

NOTE: In a special case when $r = 0$, i.e., when $b \mid a$, obviously $(a,b) = b$. □

To find the GCD of two numbers denoted by a and b, with $a > b$, according to Theorem 3.7 the problem can be reduced to finding the GCD of another pair of numbers, b and r. We can choose q and r in infinitely many ways, but only one choice leads to an efficient algorithm. To reach the end of the algorithm as quickly as possible, we must choose r as small as possible. Thus the optimal choice is r equal to the remainder of a when divided by b. We continue by applying the same procedure to finding (b,r), and so on.

We have just described the famous Euclidean algorithm. In the following, we write q_1 and r_1 instead of q and r:

$$\begin{aligned} a &= bq_1 + r_1 & 0 < r_1 < b \\ b &= r_1q_2 + r_2 & 0 < r_2 < r_1 \\ r_1 &= r_2q_3 + r_3 & 0 < r_3 < r_2 \\ r_2 &= r_3q_4 + r_4 & 0 < r_4 < r_3 \\ &\vdots \end{aligned}$$

We continue this process until we reach the first division whose remainder is 0. If $r_n = 0$, then the result we seek is r_{n-1}, the last nonzero remainder in the Euclidean algorithm. Note: Such an n exists because a descending sequence of positive integers, such as $\{r_i\}$, has a finite number of elements.

THEOREM 3.8 (EUCLIDEAN ALGORITHM). *In the Euclidean algorithm, the last nonzero remainder equals the* GCD *of the input numbers.*

PROOF: If $r_n = 0$, then r_{n-2} is divisible by r_{n-1}. Therefore their GCD is r_{n-1}. According to Theorem 3.7:

$$(a,b) = (b,r_1) = (r_1,r_2) = (r_2,r_3) = \ldots = (r_{n-2},r_{n-1}) = r_{n-1}$$

This completes the proof of correctness of the Euclidean algorithm. □

EXAMPLE 3.13. Find (543312, 65340).

SOLUTION: If we apply the Euclidean algorithm to numbers 543312 and 65340, we obtain

$$\begin{aligned} 543312 &= 8 \cdot 65340 + 20592 \\ 65340 &= 3 \cdot 20592 + 3564 \\ 20592 &= 5 \cdot 3564 + 2772 \\ 3564 &= 1 \cdot 2772 + 792 \\ 2772 &= 3 \cdot 792 + 396 \\ 792 &= 2 \cdot 396 + 0 \end{aligned}$$

This means $(543312, 65340) = 396$.

The same result follows from:

$$543312 = 2^4 \cdot 3^2 \cdot 7^3 \cdot 11 \qquad 65340 = 2^2 \cdot 3^3 \cdot 5 \cdot 11^2$$

Indeed:

$$2^2 \cdot 3^2 \cdot 11 = 396$$

□

Based on the knowledge of the two input numbers a and b only, can we estimate the number of steps required for the Euclidean algorithm to produce its output (a,b)? Yes, but before giving more details, let us first determine the two smallest numbers requiring n steps in the Euclidean algorithm.

We proceed backwards. To find the smallest of such numbers, the last division producing a nonzero remainder must have as small positive numbers as possible. Also all partial quotients except the last must be 1. Because of that, a typical step in the algorithm is:

$$r_{k-2} = r_{k-1} \cdot 1 + r_k \qquad (k = 1, 2, \ldots, n-1)$$

The last, nth step, is

$$r_{n-2} = 2r_{n-1}$$

The last quotient cannot be less than 2, otherwise the last two remainders are equal. If we add the minimum initial condition $r_{n-1} = 1$, we obtain the recursion for the Fibonacci numbers! It is easy to see that:

$$r_k = f_{n-k+1} \qquad (k = 1, 2, \ldots, n-1)$$

Therefore the numbers a and b we seek are:

$$a = f_{n+2} \qquad b = f_{n+1}$$

Thus we just proved Theorem 3.9.

THEOREM 3.9. *The smallest numbers for which the Euclidean algorithm requires n steps are the Fibonacci numbers* f_{n+2} *and* f_{n+1}.

We are now ready for Lamé's theorem from 1844.

THEOREM 3.10 (LAMÉ'S THEOREM). *The number of divisions required to determine the* GCD *of two numbers is not greater than 5 times the number of digits of the smaller number.*

PROOF: Using induction we can prove the following inequality for Fibonacci numbers:

$$f_{5k+m} > 10^k f_m$$

Assume that finding the GCD of numbers a and b, where $a > b$, requires n steps. Also assume that b has r digits in the decimal notation. According to the previous theorem, we have $b \geq f_{n+1}$. If $n > 5r$, then:

$$b \geq f_{n+1} \geq f_{5r+2} > 10^r f_2 = 10^r$$

This mean that b has $r+1$ or more digits, which contradicts the assumption that b has r digits.

Therefore if b (the smaller of the two input numbers) has r digits, the maximum number of steps is $n = 5r$. □

EXAMPLE 3.14. Lamé's upper bound is not always reached (see Example 3.13), but in the case of Fibonacci numbers, it is reached. For example for $a = f_{12} = 144$ and $b = f_{11} = 89$, according to Theorem 3.9, the number of steps is $12 - 2 = 10$. This is the same as Lamé's upper bound $5 \cdot 2 = 10$. □

We soon see that the Euclidean algorithm is not used only to find GCD's but to solve equations, too.

3.2. Important Functions in Number Theory

We determined $\tau(n)$, the number of divisors of n in Examples 2.8 and 3.11:

$$n = p_1^{\alpha_1} p_2^{\alpha_2} \dots p_r^{\alpha_r} \Rightarrow \tau(n) = (\alpha_1 + 1)(\alpha_2 + 1)\dots(\alpha_r + 1)$$

as well as $\varphi(n)$, the so called Euler's phi function, which gives the number of numbers less than n and relatively prime with it (Example 2.30):

$$n = p_1^{\alpha_1} p_2^{\alpha_2} \dots p_r^{\alpha_r} \Rightarrow \varphi(n) = n\left(1 - \frac{1}{p_1}\right)\left(1 - \frac{1}{p_2}\right)\dots\left(1 - \frac{1}{p_r}\right)$$

In this Section we define and find expressions for several other important functions often used in number theory. The property they all share is *multiplicativity*.

DEFINITION 3.7 (MULTIPLICATIVE FUNCTIONS). A function f defined over the set of integers is multiplicative if:

- $f(1) = 1$.
- $(m,n) = 1 \Rightarrow f(mn) = f(m)f(n)$.

From the expressions for $\tau(n)$ and $\varphi(n)$, we can easily check that both are multiplicative.

THEOREM 3.11. *If f_1 and f_2 are multiplicative, so is the function defined as:*

$$f(n) = f_1(n) f_2(n)$$

PROOF: We prove f satisfies the conditions from the definition of multiplicativity:

- $f(1) = f_1(1) f_2(1) = 1 \cdot 1 = 1$.
- $f(mn) = f_1(mn) f_2(mn) = f_1(m) f_1(n) f_2(m) f_2(n) = f(m) f(n)$. □

In the following, we use

$$\sum_{d|n} f(d)$$

to denote the sum of the values of f over the arguments d dividing n.

THEOREM 3.12. *If f is a multiplicative function and $n = p_1^{\alpha_1} \dots p_r^{\alpha_r}$, then:*

$$\sum_{d|n} f(d) = [1 + f(p_1) + f(p_1^2) + \dots + f(p_1^{\alpha_1})] \dots$$

$$\dots [1 + f(p_r) + f(p_r^2) + \dots + f(p_r^{\alpha_r})]$$

PROOF: If we multiply the right-hand side, we obtain the sum of terms having the following form:

$$f(p_1^{\beta_1}) \dots f(p_r^{\beta_r}) = f(p_1^{\beta_1} \dots p_r^{\beta_r})$$
$$0 \le \beta_1 \le \alpha_1, \dots, 0 \le \beta_r \le \alpha_r$$

Each term appears exactly once. Since every divisor of n has a unique canonical representation $p_1^{\beta_1} \dots p_r^{\beta_r}$, the theorem is true. It is also easy to see that if $f(n)$ is a multiplicative function, then $\sum_{d|n} f(d)$ is also a multiplicative function. □

EXAMPLE 3.15 (SUM OF DIVISORS). Function f defined as:

$$f(a) = a$$

is obviously multiplicative. The sum of all divisors of n is denoted by $\sigma(n)$ and can be written as:

$$\sigma(n) = \sum_{d|n} d$$

According to Theorem 3.12, since $f(d) = d$ is multiplicative:

$$\sigma(n) = (1 + p_1 + p_1^2 + \dots + p_1^{\alpha_1}) \dots (1 + p_r + p_r^2 + \dots + p_r^{\alpha_r})$$
$$= \frac{p_1^{\alpha_1+1} - 1}{p_1 - 1} \dots \frac{p_r^{\alpha_r+1} - 1}{p_r - 1}$$

EXAMPLE 3.16 (MÖBIUS MU FUNCTION). The Möbius mu function is defined as:

$$\mu(a) = \begin{cases} 1, & a = 1 \\ 0, & a \text{ is divisible by a square} \neq 1 \\ (-1)^r, & a \text{ is not divisible by a square} \neq 1, \text{ and } a \text{ has } r \ge 1 \text{ factors} \end{cases}$$

It is straightforward to check that the Möbius mu function is multiplicative. According to Theorem 3.11, if $f(a)$ is multiplicative, then $\mu(a)f(a)$ is also multiplicative. If $n = p_1^{\alpha_1} \dots p_r^{\alpha_r}$ then according to Theorem 3.12 and the definition of the Möbius mu function:

$$\sum_{d|n} \mu(d)f(d) = [1-f(p_1)]\dots[1-f(p_r)]$$

For example if $f(a) = 1$, we obtain the following identity:

$$\sum_{d|n} \mu(d) = \begin{cases} 0, & \text{for } a > 1 \\ 1, & \text{for } a = 1 \end{cases}$$

This is often used as the definition of the Möbius mu function.

If $f(a) = 1/a$:

$$\sum_{d|n} \frac{\mu(d)}{d} = \left(1 - \frac{1}{p_1}\right)\dots\left(1 - \frac{1}{p_r}\right)$$

It follows that:

$$\varphi(n) = \sum_{d|n} \frac{n}{d}\mu(d)$$

EXAMPLE 3.17. The following identity is very interesting:

$$\sum_{d|n} \varphi(d) = n$$

To prove it, consider $\varphi(p_k^{\beta_k}) = p_k^{\beta_k} - p_k^{\beta_k - 1}$ and Theorem 3.12.

NOTE: The identities

$$\sum_{d|n} \varphi(d) = n \qquad \varphi(n) = \sum_{d|n} \frac{n}{d}\mu(d)$$

are an example of the more general *Möbius inversion rule*:

$$\sum_{d|n} f(d) = g(n) \Leftrightarrow f(n) = \sum_{d|n} g\left(\frac{n}{d}\right)\mu(d)$$

EXAMPLE 3.18. If n has an odd number of divisors, then n is a square. Prove.

SOLUTION: If $n = p_1^{\alpha_1} p_2^{\alpha_2} \dots p_r^{\alpha_r}$, then the number of its divisors is $\tau(n) = (\alpha_1+1)(\alpha_2+1)\dots(\alpha_r+1)$. If $\tau(n)$ is odd, so is each of its factors (α_k+1) $(k = 1,2,\dots,r)$. Thus all exponents α_k are even; i.e., we can write $\alpha_k = 2\beta_k$. Now it is clear that n is a square, because:

$$n = (p_1^{\beta_1} p_2^{\beta_2} \dots p_r^{\beta_r})^2$$

3.3. Congruences

We mentioned earlier that numbers with equal remainders after dividing by the integer $m \neq 0$ have many properties in common. For that reason we introduce Definition 3.8 and the following notation.

DEFINITION 3.8 (CONGRUENCE). For integers a and b that have equal remainders after division by m, we write

$$a \equiv b \pmod{m}$$

We say that they are congruent modulo m.

This notation was introduced by Gauss in *Disquisitiones Arithmeticae*, which was published in 1801 when Gauss was only 24.

In the next few examples we examine some of the properties of numbers congruent modulo m.

EXAMPLE 3.19. If a and b are congruent modulo m, then their difference is divisible by m:

$$a \equiv b \pmod{m} \Rightarrow a = q_1 m + r \wedge b = q_2 m + r$$

Then $a - b = (q_1 - q_2)m$, therefore $m \mid (a-b)$, i.e.:

$$a - b \equiv 0 \pmod{m}$$

Note the analogy with the equals sign:

$$a \equiv b \pmod{m} \Leftrightarrow a - b \equiv 0 \pmod{m}$$

We say that the difference $(a-b)$ is congruent to zero modulo m, or shorter, that the difference $(a-b)$ is zero modulo m. Almost always we consider integers $m>0$, i.e., $m \in N$.

EXAMPLE 3.20. If:

$$a \equiv b \pmod{m}$$

then there exists q, an integer, such that:

$$a = mq + b$$

Indeed:

$$a \equiv b \pmod{m} \Rightarrow a = q_1 m + r \wedge b = q_2 m + r \Rightarrow$$
$$\Rightarrow a - b = (q_1 - q_2)m = qm$$

EXAMPLE 3.21. It is easy to verify that the relation of congruency modulo m is an equivalence relation:

- Reflexivity: Obviously $a \equiv a$.
- Symmetry: Also $a \equiv b \Rightarrow b \equiv a$.
- Transitivity: If $a \equiv b$ and $b \equiv c$, then the differences $a-b$ and $b-c$ are divisible by m. Therefore $(a-b)+(b-c) = a-c$ is also divisible by m, i.e., $a \equiv c$.

EXAMPLE 3.22. Let us use the new notation to express the fact that the value of the sum or product modulo m does not depend on the choice of numbers being added or multiplied but rather on their residues modulo m:

$$a \equiv b \pmod{m} \wedge s \equiv t \pmod{m}$$
$$\Rightarrow a + s \equiv b + t \pmod{m} \wedge as \equiv bt \pmod{m}$$

More generally if $P(x)$ is a polynomial in x with integer coefficients, then:

$$a \equiv b \pmod{m} \Rightarrow P(a) \equiv P(b) \pmod{m}$$

EXAMPLE 3.23 (CANCELLATION). If $(a,m)=1$, then:

$$ax \equiv ay \pmod{m} \Rightarrow x \equiv y \pmod{m}$$

Indeed:

$$ax \equiv ay \pmod{m} \Rightarrow m \mid a(x-y)$$

According to Example 3.10, since a and m are relatively prime, $m \mid (x-y)$, i.e., $x \equiv y \pmod{m}$.

NOTE: If $(a,m) > 1$, cancellation is not permitted. For example, $5 \cdot 4 \equiv 5 \cdot 2 \pmod{10}$, but $4 \not\equiv 2 \pmod{10}$.

EXAMPLE 3.24. If $(a,m) = d \geq 1$, then:

$$ax \equiv ay \pmod{m} \Rightarrow x \equiv y \left(\bmod \frac{m}{d}\right)$$

As in Example 3.23:

$$ax \equiv ay \pmod{m} \Rightarrow m \mid a(x-y) \Rightarrow \frac{m}{d} \,\Big|\, \frac{a}{d}(x-y)$$

Finally:

$$\left(\frac{a}{d}, \frac{m}{d}\right) = 1 \Rightarrow \frac{m}{d} \mid (x-y) \Rightarrow x \equiv y \left(\bmod \frac{m}{d}\right)$$

EXAMPLE 3.25. If $a \equiv b \pmod{qm}$, then $a \equiv b \pmod{m}$. Indeed:

$$a \equiv b \pmod{qm} \Rightarrow qm \mid (a-b) \Rightarrow m \mid (a-b) \Rightarrow a \equiv b \pmod{m}$$

EXAMPLE 3.26. What can we conclude if we know that:

$$x \equiv y \pmod{m} \qquad x \equiv y \pmod{n}?$$

From $m \mid (x-y)$ and $n \mid (x-y)$, we find that $(x-y)$ is a common multiple of m and n; therefore $[m,n] \mid (x-y)$, i.e., $x \equiv y \pmod{[m,n]}$.

From Example 3.25 we see that the converse is also true, so:

$$x \equiv y \pmod{m} \wedge x \equiv y \pmod{n} \Leftrightarrow x \equiv y \pmod{[m,n]} \qquad \square$$

Since the congruence is an equivalence relation, the classes of congruence modulo m:

$$[0],\ [1],\ \ldots,\ [m-1]$$

form a complete partition of the set of integers. In other words, the classes of residues modulo m are disjoint sets, and their union is the set of integers Z.

The set $\{0,1,\ldots,(m-1)\}$ has a special name, the *complete residue system* modulo m. In fact any set in which every residue class has exactly one representative is called a complete residue system modulo m.

To prove the next few theorems, we need another residue system, the so called *reduced residue system* modulo m. It is constructed from the complete residue system modulo m by eliminating all numbers not relatively prime with m. Obviously the number of elements in the reduced residue system modulo m equals Euler's phi function $\varphi(m)$.

EXAMPLE 3.27. If $m=9$, the most often used complete residue system is the following set: $\{0,1,2,3,4,5,6,7,8\}$. The most often used reduced residue system is $\{1,2,4,5,7,8\}$. The sets $\{0,1,2,3,4,5,6,7,17\}$ and $\{1,2,4,5,7,17\}$ are also complete and reduced residue systems, respectively.

Let us consider Theorem 3.13, which we use to prove Euler's and Wilson's theorems.

THEOREM 3.13. *If $\{x_1,x_2,\ldots,x_{\varphi(m)}\}$ is a reduced residue system modulo m and $(a,m)=1$, then the set $\{ax_1,ax_2,\ldots,ax_{\varphi(m)}\}$ is a reduced residue system modulo m, too. (Note: The order of the elements in a set is not important.)*

PROOF: Suppose $\{ax_1,ax_2,\ldots,ax_{\varphi(m)}\}$ is not a reduced residue system modulo m. Then for some $i\neq j$, $ax_i\equiv ax_j \pmod m$, or for some i, $(ax_i,m)\neq 1$. But since $(a,m)=1$, in the former case $x_i\equiv x_j \pmod m$; in the latter case $(ax_i,m)\neq 1$. Both cases imply that $\{x_1,x_2,\ldots,x_{\varphi(m)}\}$ is not a reduced residue system modulo m either. Note:

$$\begin{matrix}(a,m)=1 \wedge (x_i,m)=1 \\ (i=0,1,\ldots,\varphi(m))\end{matrix} \quad\Rightarrow\quad \begin{matrix}(ax_i,m)=1 \\ (i=0,1,\ldots,\varphi(m))\end{matrix}$$

This completes the proof. □

The proof of Theorem 3.14 is completely analogous.

THEOREM 3.14. *If* $\{x_1, x_2, \ldots, x_m\}$ *is a complete residue system modulo m and* $(a,m) = 1$, *then the set* $\{ax_1, ax_2, \ldots, ax_m\}$ *is a complete residue system modulo m, too. (Note: The order of the elements in a set is not important.)*

We are now ready to proceed with the proof of Euler's theorem (Theorem 3.15).

THEOREM 3.15 (EULER'S THEOREM). *If* $(a,m) = 1$, *then:*

$$a^{\varphi(m)} \equiv 1 \pmod{m}$$

PROOF: According to Theorem 3.14, to every number from the reduced residue system $\{x_1, x_2, \ldots, x_{\varphi(m)}\}$, there corresponds exactly one element of the reduced residue system $\{ax_1, ax_2, \ldots, ax_{\varphi(m)}\}$ such that:

$$x_i \equiv ax_j \pmod{m}$$

Therefore we can write

$$ax_1 ax_2 \ldots ax_{\varphi(m)} \equiv x_1 x_2 \ldots x_{\varphi(m)} \pmod{m}$$

Since $(x_i, m) = 1$ $[i = 0, 1, \ldots, \varphi(m)]$, we are allowed to cancel the x_is. Thus:

$$a^{\varphi(m)} \equiv 1 \pmod{m}. \qquad \square$$

Fermat's lesser theorem from 1640 (Theorem 3.16) is a special case of Euler's theorem.

THEOREM 3.16 (FERMAT'S LESSER THEOREM). *If p is a prime and a is an integer not divisible by p, then:*

$$a^{p-1} \equiv 1 \pmod{p}$$

PROOF: The proof reduces to the observation that if p is a prime, then $\varphi(p) = p - 1$. Two different proofs can be found in Examples 2.58 and 3.28. □

NOTE: Around 500 B.C., Chinese mathematicians knew that if p is a prime, then $2^p \equiv 2 \pmod{p}$. This is a special case of Fermat's lesser theorem. But they, and also Leibniz centuries after them, thought the converse was true too; i.e., if $2^n \equiv 2 \pmod{n}$, then n must be prime. The smallest counterexample to their belief is $n = 341 = 11 \cdot 31$, for which:

$$2^{341} \equiv \left(2^{10}\right)^{34} \cdot 2 \equiv 1^{34} \cdot 2 \equiv 2 \pmod{341}$$

Such numbers as 341 are called *pseudoprimes* to the base 2. In Example 3.45 we show that there are infinitely many pseudoprimes to the base 2; there are pseudoprimes to other bases as well. Some numbers are pseudoprime for all bases. Such numbers are called *absolute pseudoprimes*, or *Carmichael numbers*. The smallest such number is 561. □

Theorem 3.17 proves a formula that holds for all primes and only for primes. Unfortunately that formula is not very useful for primality testing, because it requires considerable computation. Theorem 3.17 was discovered by Leibniz in 1682, but it is better known today as Wilson's theorem.

THEOREM 3.17 (WILSON'S THEOREM). *Congruence:*

$$(p-1)! \equiv -1 \pmod{p}$$

is true if and only if p is a prime.

PROOF: If $(p-1)! \equiv -1 \pmod{p}$, then if p is not a prime, there exists a prime $q < p$ that divides p. But since $q < p$, q is a factor in $(p-1)!$; therefore $(p-1)!+1$ cannot be divisible by p. Therefore p must be prime if $(p-1)! \equiv -1 \pmod{p}$ is true.

On the other hand for p prime, consider

$$(p-1)! = 1 \cdot 2 \cdot 3 \cdot \ldots \cdot (p-2) \cdot (p-1)$$

Obviously $1 \equiv 1 \pmod{p}$ and $(p-1) \equiv -1 \pmod{p}$. For the remaining terms in the product, for any r, $2 \leq r \leq (p-2)$, there exists exactly one $s \neq r$, $2 \leq s \leq (p-2)$, such that $rs \equiv 1 \pmod{p}$.

Since $\{1, 2, \ldots, (p-1)\}$ is a reduced residue system modulo p, so is $\{r, 2r, \ldots, (p-1)r\}$. Therefore exactly one element of the latter set is 1 modulo p. Since $2 \leq r \leq (p-2)$, it is not r. It is also not $r(p-1)$, because $r(p-1) \equiv rp - r \equiv -r \pmod{p}$. Furthermore it is not $r \cdot r$, otherwise $r^2 \equiv 1$

(mod p), which implies $(r-1)(r+1) \equiv 0 \pmod{p}$; i.e., $r \equiv 1 \pmod{p}$ or $r \equiv -1 \pmod{p}$, which contradicts the initial assumption $2 \leq r \leq (p-2)$. This completes the proof. □

EXAMPLE 3.28 (AGAIN FERMAT'S LESSER THEOREM). Prove the following:

1. Product of two consecutive integers is divisible by 2.
2. Product of three consecutive integers is divisible by 6.
3. Product $n(n-1)(n-2)\ldots(n-k+1)$ is divisible by $k!$

SOLUTION:

1. One of the two consecutive integers is even. Therefore their product is even too.
2. Among the three consecutive integers, one is divisible by 3. In addition at least one of these is even. Therefore their product is divisible by $3 \cdot 2 = 6$.
3. In Chapter 2 we saw that the number of subsets with k elements picked out of an n-element set equals

$$\frac{n(n-1)(n-2)\ldots(n-k+1)}{k!} = \frac{n!}{k!\,(n-k)!}$$

Therefore it must be an integer.

Another way of showing that follows. First note that the exponent of an arbitrary prime number q in the canonical decomposition of $m!$ is

$$\lfloor m/q \rfloor + \lfloor m/q^2 \rfloor + \lfloor m/q^3 \rfloor + \ldots$$

because among the integers $\leq m$, $\lfloor m/q \rfloor$ are divisible by q, $\lfloor m/q^2 \rfloor$ are divisible by q^2, etc.

To show that

$$\frac{n!}{k!\,(n-k)!}$$

is an integer, we apply the previous finding to numbers $n!$, $k!$, and $(n-k)!$, noting that $\lfloor a+b \rfloor \geq \lfloor a \rfloor + \lfloor b \rfloor$.

NOTE: If $n = p$, where p is any prime number, then for $2 \leq k \leq (p-1)$:

$$\binom{p}{k} = \frac{p(p-1)(p-2)\dots(p-k+1)}{k!}$$

is divisible by p, because p is a prime and $p > k$, so p is not a factor in $k!$.

Therefore the binomial theorem modulo a prime becomes

$$(a+b)^p \equiv a^p + b^p \pmod{p}$$

The result can easily be extended to the powers of sums of three or more terms:

$$(a_1 + a_2 + \dots + a_r)^p \equiv a_1^p + a_2^p + \dots + a_r^p \pmod{p}$$

If we let $a_1 = a_2 = \dots = a_r = 1$, we obtain Fermat's lesser theorem:

$$r^p \equiv r \pmod{p}$$

If Fermat's lesser theorem is written in this form, it does not require the condition $(r,p) = 1$.

EXAMPLE 3.29. If the number 1968 is written as a sum of several integers and these integers are cubed and added together, we obtain the number divisible by 6. Prove.

SOLUTION: Instead of 1968, we write n to prove the more general statement that the sum of the cubes of any partition of n is congruent with n modulo 6. If $n = n_1 + \dots + n_r$, then:

$$\begin{aligned} n_1^3 + \dots + n_r^3 &= (n_1^3 - n_1) + \dots + (n_r^3 - n_r) + n \\ &= n_1(n_1 - 1)(n_1 + 1) + \dots + n_r(n_r - 1)(n_r + 1) + n \end{aligned}$$

Since each of the numbers $n_k(n_k - 1)(n_k + 1)$ is divisible by 6:

$$n_1^3 + \dots + n_r^3 \equiv n \pmod{6}$$

NOTE: r, the number of terms in the partition of n, does not play any role at all.

EXAMPLE 3.30. Find the last digit of 777^{333}.

SOLUTION: For any n the power n^k ends with the same digit as r^k, where r is the last digit of n. Indeed if $n = 10q + r$, then:

$$n^k = (10q + r)^k = 10q' + r^k \equiv r^k \pmod{10}$$

Thus we are interested in finding the last digit of 7^{333}. Since:

$$7^0 \equiv 1 \qquad 7^1 \equiv 7 \qquad 7^2 \equiv 9 \qquad 7^3 \equiv 3 \qquad 7^4 \equiv 1 \pmod{10} \qquad \ldots$$

we see that the last digits are periodically repeated, with the period equal to 4. Since $333 \equiv 1 \pmod 4$, $777^{333} \equiv 7 \pmod{10}$.

EXAMPLE 3.31. If $n > 4$ is a composite number, then $n \mid (n-1)!$ Prove.

SOLUTION: If n is a composite number, then there exist integers n_1 and n_2 such that $n = n_1 n_2$ and $n_1, n_2 > 1$. Obviously $n_1, n_2 < n$. Two cases are possible:

- $n_1 \neq n_2$. Then both numbers enter the product $(n-1)! = 1 \cdot 2 \cdot \ldots \cdot (n-1)$ and obviously $n \mid (n-1)!$
- $n_1 = n_2$. Then $n = n_1^2$, and since $n > 4$, $n_1 > 2$. Hence $n = n_1^2 > 2n_1$. This implies that n_1 and $2n_1$ are less than n and thus enter the product $(n-1)! = 1 \cdot 2 \cdot \ldots \cdot (n-1)$. Therefore we find $n \mid (n-1)!$

EXAMPLE 3.32 (PASCAL'S CRITERION). If the digits of a are $a_n, a_{n-1}, \ldots, a_1$, and a_0, or in other words if:

$$a = 10^n a_n + 10^{n-1} a_{n-1} + \ldots + 10 a_1 + a_0 = \overline{a_n a_{n-1} \ldots a_1 a_0}$$

then it is straightforward to prove the general divisibility criterion, due to Pascal:

$$m \mid a \Leftrightarrow m \mid (r_n a_n + r_{n-1} a_{n-1} + \ldots + r_1 a_1 + r_0 a_0)$$

where $10^k \equiv r_k \pmod m \quad (k = 0, 1, 2, \ldots, n)$.

Using this criterion we can find the divisibility criteria for powers of 2 and 5 and for numbers 3, 9, 7, 11, and 13:

- Divisibility by 2^r $(r \in N)$: A number is divisible by 2^r if and only if the number formed from its last r digits is divisible by 2^r.
- Divisibility by 5^r $(r \in N)$: A number is divisible by 5^r if and only if the number formed from its last r digits is divisible by 5^r.
- Divisibility by 3: A number is divisible by 3 if and only if the sum of its digits is divisible by 3.
- Divisibility by 9: A number is divisible by 9 if and only if the sum of its digits is divisible by 9.
- Divisibility by 11: There are at least three criteria for 11:

$$\begin{aligned} 11 \mid a &\Leftrightarrow 11 \mid (a_0 - a_1 + a_2 - a_3 + a_4 - \dots) \\ &\Leftrightarrow 11 \mid (\overline{a_1 a_0} + \overline{a_3 a_2} + \overline{a_5 a_4} + \overline{a_7 a_6} + \dots) \\ &\Leftrightarrow 11 \mid (\overline{a_2 a_1 a_0} - \overline{a_5 a_4 a_3} + \overline{a_8 a_7 a_6} - \dots) \end{aligned}$$

- Divisibility by 7 and 13: Since $7 \cdot 11 \cdot 13 = 1001$, the criteria for 7 and 13 are the same as the third criterion for 11.

3.4. Diophantine Equations

Any equation whose solutions are required to be integers or rational numbers is called a *Diophantine equation*. Only some of these equations can be solved systematically, i.e., using some algorithm. For others we usually need to know a lot of number theory and have a lot of ideas and imagination.

Linear Congruences with One Unknown

The simplest Diophantine equations are

$$ax \equiv c \pmod{m} \tag{3.2}$$

That is:

$$ax + my = c \tag{3.3}$$

According to Example 3.20, Eqs. (3.2) and (3.3) are equivalent.

Let us first consider the existence of the solution of Eq. (3.2) or Eq. (3.3). Recall Euler's theorem:

$$(a,m) = 1 \Rightarrow a^{\varphi(m)} \equiv 1 \pmod{m}$$

According to Euler's theorem, if $(a,m) = 1$, the solution exists, and furthermore we can write the following explicit* formula for it:

$$x \equiv ca^{\varphi(m)-1} \pmod{m}$$

We see that there are infinitely many solutions — all numbers from the same residue class modulo m as $ca^{\varphi(m)-1}$.

In the case when $(a,m) = d > 1$, there are two possibilities.

Case 1: When $d \mid c$, we cancel d on both sides of the congruence to obtain the new equation:

$$\frac{a}{d}x \equiv \frac{c}{d} \left(\mathrm{mod}\ \frac{m}{d}\right)$$

where:

$$\left(\frac{a}{d}, \frac{m}{d}\right) = 1$$

Thus we reduced the problem to the previously encountered and solved case.

Case 2: When $d \nmid c$, there is no solution because if $d \mid a$ and $d \mid m$, then $d \mid (ax + my) = c$.

EXAMPLE 3.33. The equation:

$$15x + 25y = 14$$

does not have integer solutions because 14 is not divisible by 5.

The equation:

$$15x + 25y = 10$$

has solutions, which can be found from the simpler equation obtained by dividing by 5:

$$3x + 5y = 2$$

---

*Unfortunately the application of this formula requires a lot of calculations. We soon learn how to use the Euclidean algorithm, which is computationally less involved.

One solution is evident:

$$x_0 = -1 \qquad y_0 = 1$$

Now it is easy to see that the other solutions are given by:

$$x = x_0 + 5k \qquad y = y_0 - 3k \qquad (k \in Z)$$

because the terms $3 \cdot 5k$, and $5 \cdot (-3k)$ add up to zero, leaving only $3x_0 + 5y_0 = 2$.

Let us see if Euler's theorem gives the same solution:

$$x \equiv ca^{\varphi(m)-1} \equiv 2 \cdot 3^{\varphi(5)-1} \equiv 2 \cdot 3^{4-1} \equiv 54 \equiv 4 \equiv -1 \pmod 5 \qquad \square$$

Using intuition or Euler's phi function is not satisfactory even for small values of a, m, and c. To show how the Euclidean algorithm can be used to solve linear congruences in one unknown, we need Theorem 3.18:

THEOREM 3.18. *If $d = (a,b)$, then there exist integers α and β such that:*

$$\alpha a + \beta b = d$$

PROOF: Among all integers of the form $xa + yb$ (among these are both positive and negative integers, even 0), there exists *the smallest positive* such number. Denote it as $n = \alpha a + \beta b$. First we prove that n is a common divisor of a and b, and then that $n = d = (a,b)$.

If n is not a divisor of a, according to Theorem 3.2, we would be able to find a unique pair of numbers q and r, where $0 < r < n$, such that $a = nq + r$. Then:

$$r = a - nq = a - (\alpha a + \beta b)q = (1 - \alpha q)a - \beta q b = x'a + y'b$$

Hence r is one of the numbers $xa + yb$. But since $0 < r < n$, it is positive and smaller than *the smallest positive* such number, a contradiction. Thus a must be divisible by n. Similarly we can prove that $n \mid b$.

Since $d = (a,b)$, we have

$$a = vd, \; b = wd$$

for some integers v and w. Then:

$$n = \alpha a + \beta b = \alpha v d + \beta w d = (\alpha v + \beta w)d$$

We see that $d \mid n$, which implies $n \geq d$. But d is the *greatest* common divisor of a and b; hence the only possibility is $n = d$. This completes the proof. □

Why do we need Theorem 3.18? If we know how to find α and β such that:

$$\alpha a + \beta m = d \qquad (a, m) = d$$

(we just proved they exist), we know how to find the initial solutions x_0 and y_0 of the equation:

$$ax + my = c$$

where $d \mid c$.

It is easy to verify that:

$$x_0 = \frac{c}{d}\alpha \qquad y_0 = \frac{c}{d}\beta$$

can be used as the initial solutions.

To solve the equation:

$$ax + my = c$$

we must determine $d = (a, m)$ to see if the equation has solutions. The most efficient method of doing that is the Euclidean algorithm. Note: We can use the intermediate results of the Euclidean algorithm to find the numbers α and β. Consider Example 3.34:

EXAMPLE 3.34. Earlier from the following, we found that $(543312, 65340) = 396$ as follows:

$$543312 = 8 \cdot 65340 + 20592$$
$$65340 = 3 \cdot 20592 + 3564$$
$$20592 = 5 \cdot 3564 + 2772$$
$$3564 = 1 \cdot 2772 + 792$$
$$2772 = 3 \cdot 792 + 396$$
$$792 = 2 \cdot 396 + 0$$

Now starting from $2772 - 3 \cdot 792 = 396$, and substituting $792 = 3564 - 1 \cdot 2772$, we find $-3 \cdot 3564 + 4 \cdot 2772 = 396$. We continue to ascend until we reach the initial numbers 543312 and 65340. Then we find that:

$$73 \cdot 543312 + (-607) \cdot 65340 = 396$$

Thus to solve the equation:

$$543312x + 65340y = 1188$$

we first use the Euclidean algorithm to find that $(543312, 65340) = 396$, then we verify that $396 \mid 1188$, and continue with the *extended Euclidean algorithm*, until we find that α and β are 73 and -607, respectively.

Finally we find

$$x_0 = \frac{1188}{396} \cdot 73 = 219 \qquad y_0 = \frac{1188}{396} \cdot (-607) = -1821$$

This implies that all solutions of the initial equation are

$$x = 219 + \frac{65340}{396}k \qquad y = -1821 - \frac{543312}{396}k \quad (k \in Z)$$

Since $65340/396 = 165$ and $543312/396 = 1372$, we can also take $x_0 = 54$ and $y_0 = -449$ as the initial solutions. (Also see Example 3.41). □

Especially useful notation for solving linear congruences in one unknown using the extended Euclidean algorithm includes the so-called *continued fractions*, for example:

$$\frac{a}{m} = q_1 + \frac{1}{q_2 + \frac{1}{\cdots + \frac{1}{q_n}}}$$

Continued fractions deserve much more attention than we can afford to give them here. For example the representation of the golden section $\phi = (1 + \sqrt{5})/2 = 1.6180\ldots$ using continued fractions is most interesting. Since ϕ is an irrational number, the corresponding continued fraction is of an infinite order. Since $\phi = 1 + (1/\phi)$:

$$\phi = 1 + \frac{1}{1 + \frac{1}{1 + \cdots}}$$

Chinese Remainder Theorem

In the first century A.D. the Chinese mathematician Sun-Tsu solved the problem that can be reduced to finding the integers x that are congruent to 2, 3, and 2 modulo 3, 5, and 7, respectively:

$$\left.\begin{array}{l} x \equiv 2 \pmod{3} \\ x \equiv 3 \pmod{5} \\ x \equiv 2 \pmod{7} \end{array}\right\} \Rightarrow x \equiv 23 \pmod{105}$$

Although the first general solution for problems of this type was given by Euler, Theorem 3.19 is usually called the *Chinese remainder theorem*.

THEOREM 3.19 (CHINESE REMAINDER THEOREM). *Let*

$$n = n_1 \dots n_k$$

where $n_1, \dots, n_k$ are relatively prime, i.e.:

$$(n_i, n_j) = 1 \qquad (i \neq j)$$

Then for any given ordered k-tuple $(a_1, \dots, a_k)$, there is a unique number a modulo n such that:

$$a \equiv a_1 \pmod{n_1}$$

$$\vdots$$

$$a \equiv a_k \pmod{n_k}$$

PROOF: Let the numbers $m_i \quad (i = 1, \dots, k)$ be such that $m_i = n/n_i$ and let the numbers M_i $(i = 1, \dots, k)$ be such that:

$$m_i M_i \equiv 1 \pmod{n_i} \qquad (i = 1, \dots, k)$$

Then:

$$a \equiv m_1 M_1 a_1 + \dots + m_k M_k a_k \pmod{n}. \qquad \square$$

The last formula solves problems similar to the one considered by Sun-Tsu. Numbers M_i are usually determined using the extended Euclidean algorithm.

NOTE: Some fast calculation methods are based on this theorem and the fact that the k-tuple corresponding to the sum $(a+b) \pmod{n}$ is

$$((a_1+b_1) \pmod{n_1}, \ldots, (a_k+b_k) \pmod{n_k})$$

while the k-tuple corresponding to the product $(ab) \pmod{n}$ is

$$((a_1 b_1) \pmod{n_1}, \ldots, (a_k b_k) \pmod{n_k})$$

EXAMPLE 3.35. Find four consecutive integers such that each of them is divisible by a square > 1.

SOLUTION: The simplest case can be written as a system of four equations:

$$\begin{aligned} a &\equiv 0 \pmod{2^2} \\ a+1 &\equiv 0 \pmod{3^2} \\ a+2 &\equiv 0 \pmod{5^2} \\ a+3 &\equiv 0 \pmod{7^2} \end{aligned} \quad \text{i.e.,} \quad \begin{aligned} a &\equiv 0 \pmod{4} \\ a &\equiv 8 \pmod{9} \\ a &\equiv 23 \pmod{25} \\ a &\equiv 46 \pmod{49}. \end{aligned}$$

We chose the squares of primes to satisfy the conditions of the Chinese remainder theorem. Now:

$$\begin{array}{llll} a_1 = 0 & n_1 = 4 & m_1 = 11025 & M_1 = 1^{-1} \equiv 1 \pmod{4} \\ a_2 = 8 & n_2 = 9 & m_2 = 4900 & M_2 = 4^{-1} \equiv 7 \pmod{9} \\ a_3 = 23 & n_3 = 25 & m_3 = 1764 & M_3 = 14^{-1} \equiv 9 \pmod{25} \\ a_4 = 46 & n_4 = 49 & m_4 = 900 & M_4 = 18^{-1} \equiv 30 \pmod{49} \\ & n\ = 44100 & & \end{array}$$

and:

$$\begin{aligned} a &= 11025 \cdot 1 \cdot 0 + 4900 \cdot 7 \cdot 8 + 1764 \cdot 9 \cdot 23 + 900 \cdot 30 \cdot 46 \\ &= 1881548 \equiv 29348 \pmod{44100} \end{aligned}$$

Hence the numbers 29348, 29349, 29350, and 29351, which are divisible by 2^2, 3^2, 5^2, and 7^2, respectively, are one possible answer.

Pythagorean Triples

At the beginning of Chapter 3 we mentioned that during the whole millennium before Pythagoras, Babylonian mathematicians knew a systematic way of determining the triples of integers representing the sides of right-angled triangles, i.e., satisfying the Pythagorean theorem:

$$a^2 + b^2 = c^2$$

The simplest such triple is $(3,4,5)$, which is sometimes erroneously called the *Egyptian triangle.* (Apparently builders of the Egyptian pyramids used to construct the right angle by constructing the triangle with sides 3,4, and 5.)

If (a,b,c) is a Pythagorean triple and the numbers a,b, and c are relatively prime, then we call (a,b,c) a *primitive triple.* If we find all primitive triples, then we can find all others, too, by multiplying a,b, and c by the same factors.

If (a,b,c) is a primitive Pythagorean triple, then one of the numbers a or b is even and the other is odd, implying that c must be odd. If both a and b are even, then c must be even, too, so the triple is not primitive. If a and b were both odd, then:

$$c^2 = a^2 + b^2 \equiv 1 + 1 \equiv 2 \pmod 4$$

But the squares of integers can be only 0 or 1 modulo 4.

Let a be even; i.e., let $a = 2\alpha$. Then:

$$a^2 = 4\alpha^2 = c^2 - b^2 = (c-b)(c+b)$$

The numbers $(c-b)$ and $(c+b)$ are even, so we can divide the preceding equation by 4 to obtain

$$\alpha^2 = \frac{c-b}{2} \cdot \frac{c+b}{2}$$

The terms on the right-hand side are relatively prime [otherwise their common divisor must divide both their sum and difference, i.e., numbers b and c and therefore a as well; hence (a,b,c) are not primitive]; therefore they must be squares themselves:

$$\frac{c-b}{2} = m^2 \qquad \frac{c+b}{2} = n^2$$

Finally, $a = 2mn$, $b = m^2 - n^2$, and $c = m^2 + n^2$, where m and n are arbitrary relatively prime integers, and $m > n$.

Since for every $m,n \in Z$ $(2mn)^2 + (m^2 - n^2)^2 = (m^2 + n^2)^2$ we have just proved the following

THEOREM 3.20 (PYTHAGOREAN TRIPLES). *All primitive Pythagorean triples are given by:*

$$a = 2mn \qquad b = m^2 - n^2 \qquad c = m^2 + n^2$$

where $m, n \in N$, $(m,n) = 1$, *and* $m > n$.

Fermat's Last Theorem

During the Dark Ages, learned people in Europe were unaware of the great mathematical discoveries that had been made by Ancient Greek mathematicians. The same was true of other sciences, medicine, and philosophy. During the High Middle Ages and Renaissance the teachings of the Ancient Greeks slowly returned to Europe, mostly through Arab translations, although many of the original Greek books were preserved in great European libraries. A real discovery for the fifteenth-century mathematicians was the Diophantus' *Arithmetica*, which considered problems reducible to equations whose solutions were required to be integers or rational numbers, hence the name *Diophantine equations*. The first translations into Latin, the scientific language of that time, appeared in the sixteenth century.

Besides its role in introducing European mathematicians to the ancient knowledge about numbers, Diophantus' *Arithmetica* played another very important role in the history of mathematics. While reading this work, the French mathematician Pierre de Fermat* used to write comments in the margins, among them many important theorems and hypotheses. In 1670 Fermat's son published Diophantus' *Arithmetica* with his father's comments, which encouraged mathematicians of later times to try to prove or disprove Fermat's hypotheses. Through these attempts many new fields of mathematics were discovered.

Especially difficult and challenging was the so-called *Fermat's last theorem*, which was proved only recently.†

THEOREM 3.21 (FERMAT'S LAST THEOREM). *There are no integers* x, y, z, *nor* n

*Fermat was a contemporary of Descartes and Pascal and together with them a founder of such disciplines as analytic geometry and probability. He contributed a lot to physics too, e.g., *Fermat's principle* in optics.

†Unlike many previous proofs that sooner or later were found to be incorrect, the proof by Andrew Wiles is accepted as a serious and successful attempt. Nevertheless due to the specialized knowledge required to understand it, many years will pass before the last sceptics accept it as the final proof.

such that for $n > 2$:

$$x^n + y^n = z^n$$

The proof of Fermat's last theorem is very complicated and lengthy, and certainly unsuitable for this book. We say only that it was announced by the Princeton mathematician Andrew Wiles in 1993 and published in the *Annals of Mathematics* in May 1995. (See Ref. [64].)

In the margins of the Diophantus' book Fermat wrote:

> It is impossible to separate a cube into two cubes, or a biquadrate into two biquadrates, or in general any power higher than the second into powers of like degree; I have discovered a truly remarkable proof which this margin is too small to contain.

3.5. Problems

EXAMPLE 3.36. Prove that $\forall n \in N$ the number $A = n^5 - n$ is divisible by 30.

SOLUTION: From $A = n^5 - n = n(n-1)(n+1)(n^2+1)$, we see that $6 \mid A$ because n, $(n-1)$, and $(n+1)$ are three consecutive numbers. To complete the proof, it remains to show $5 \mid A$.

If none of the numbers n, $(n-1)$, or $(n+1)$ is divisible by 5, then n is congruent to 2 or 3 modulo 5, i.e., $n = 5k \pm 2$, where k is some integer. Then:

$$n^2 + 1 = (5k \pm 2)^2 + 1 = 25k^2 \pm 20k + 5 = 5k'$$

Thus $30 \mid (n^5 - n)$.

EXAMPLE 3.37 (DISTRIBUTION OF PRIMES). The number of primes $\leq x$ is denoted by $\pi(x)$. Around 1846 Chebyshev showed that $0.92x/(\ln x) < \pi(x) < 1.11x/(\ln x)$, and if there exists the limit

$$\lim_{x \to \infty} \frac{\pi(x)}{x/\ln x}$$

then it must equal 1. In 1896 independent of one another, Hadamard and de la Valee–Poussin proved the existence of that limit and hence the asymptotic formula for $\pi(x)$:

$$\pi(x) \sim \frac{x}{\ln x}$$

or, in other words:

$$\lim_{x \to \infty} \frac{\pi(x)}{x/\ln x} = 1$$

EXAMPLE 3.38. Find all natural numbers a, b, and c satisfying:

$$\frac{1}{ab} + \frac{1}{ac} + \frac{1}{bc} = 1$$

SOLUTION: If the sum of three positive numbers is 1, then at least one of them is $\geq 1/3$, and all of them are < 1. Let

$$\frac{1}{ab} \geq \frac{1}{3}$$

Then $1 < ab \leq 3$. If we check all possible cases, we find that all solutions are permutations of the set $\{1,2,3\}$.

EXAMPLE 3.39. Solve the following equation in the set of natural numbers:

$$1! + 2! + 3! + \ldots + x! = y^2$$

SOLUTION: We prove that this equation has only two solutions:

$$x = 1 \quad y = 1 \qquad \text{and} \qquad x = 3 \quad y = 3$$

It is easy to verify that for $x = 1, 2$, and 3 the only solutions are $(1,1)$ and $(3,3)$. When $x = 4$ the equation does not have a solution because 33 is not a square.

When $x \geq 5$ the last digit of $x!$ is 0, so the sum on the left-hand side always ends in 3. On the other hand, squares of integers end in $0, 1, 4, 5, 6, 9$, i.e., never in 3.

EXAMPLE 3.40. Are there any solutions to:

$$x!y! = z!$$

if they are required to be integers > 5?

SOLUTION: This equation has infinitely many solutions of the following form:

$$x = n \quad y = (n! - 1) \quad z = n!$$

where $n \in N$ and $n > 5$.

There is also at least one solution that is not of that form: $x = 6$, $y = 7$, $z = 10$.

EXAMPLE 3.41 (EULER'S METHOD). Euler's method for solving linear Diophantine equations represents an elegant transition from intuitive solutions to a method based on the extended Euclidean algorithm and continued fractions. Consider our earlier example:

$$543312x + 65340y = 1188 \Rightarrow y = -8x + \underbrace{\frac{-20592x + 1188}{65340}}_{t}$$

We see that t must be an integer, which gives us an auxiliary Diophantine equation:

$$65340t = -20592x + 1188$$

In a similar fashion we obtain the following equations with ever smaller coefficients:

$$\begin{aligned} 20592u &= -3564t + 1188 \\ 3564v &= -2772u + 1188 \\ 2772w &= -792v + 1188 \\ 792z &= -396w + 396 \end{aligned}$$

i.e., $2z = -w + 1$.

If we set $z = 0$, then $w_0 = 1$, $v_0 = -2$, $u_0 = 3$, $t_0 = -17$, and the initial solutions are $x_0 = 54$ and $y_0 = -449$.

If z is left undetermined, then the general solution (see also Example 3.34) is:

$$x = 54 + 165z \qquad y = -449 - 1372z \quad (z \in Z)$$

EXAMPLE 3.42 (GAUSS'S METHOD). We can use Example 3.23 and other properties of congruences to solve $27x + 100y = 1$ as Gauss did

$$27x \equiv 1 \pmod{100} \Rightarrow$$
$$\Rightarrow \equiv \frac{1}{27} \equiv \frac{-99}{27} \equiv \frac{-11}{3} \equiv \frac{-111}{3} \equiv -37 \equiv 63 \pmod{100}$$

EXAMPLE 3.43 (FIBONACCI NUMBERS). The following identities involving the sequence of Fibonacci numbers can be easily proved using mathematical induction:

$$f_1 + f_2 + f_3 + \ldots + f_n = f_{n+2} - 1$$
$$f_1 + f_3 + f_5 + \ldots + f_{2n-1} = f_{2n}$$
$$f_2 + f_4 + f_6 + \ldots + f_{2n} = f_{2n+1} - 1$$
$$f_1 f_2 + f_2 f_3 + f_3 f_4 + \ldots + f_{2n-1} f_{2n} = f_{2n}^2$$
$$f_1 f_2 + f_2 f_3 + f_3 f_4 + \ldots + f_{2n} f_{2n+1} = f_{2n+1}^2 - 1$$
$$f_1^2 + f_2^2 + f_3^2 + \ldots + f_n^2 = f_n f_{n+1}$$
$$\binom{n}{1} f_1 + \binom{n}{2} f_2 + \binom{n}{2} f_3 + \ldots + \binom{n}{n} f_n = f_{2n}$$
$$f_{n-1} f_{n+1} - f_n^2 = (-1)^n$$

NOTE: The last identity was discovered by Cassini in 1680. It can also be proved using the result of Problem A.12 in Appendix A.

EXAMPLE 3.44 (FIBONACCI NUMBERS AND DIVISIBILITY). Mathematical induction can be used to prove

$$f_{n+k} = f_k f_{n+1} + f_{k-1} f_n$$

If we set $k = n, 2n, 3n$, etc., we find that (proof is again by induction) $f_n \mid f_{mn}$, i.e., if $r \mid s$, then $f_r \mid f_s$. In general it can be shown that $(f_m, f_n) = f_{(m,n)}$; hence we have the converse, too: If $f_r \mid f_s$ then $r \mid s$, hence $f_r \mid f_s$ if and only if $r \mid s$.

EXAMPLE 3.45 (PSEUDOPRIMES). The note following Theorem 3.16 defines pseudoprimes to the base 2 as all composite integers n for which $2^n \equiv 2 \pmod{n}$. We showed that 341 is one such number, and we now show that there are infinitely many pseudoprimes. Indeed we show that if n is an odd pseudoprime, then so is $m = 2^n - 1$. This construction of pseudoprimes can be continued *ad infinitum*.

Since n is composite, we can write $n = uv$, with $1 < u, v < n$; hence the following number is composite as well:

$$m = 2^n - 1 = 2^{uv} - 1 = (2^u - 1)(2^{u(v-1)} + \ldots + 1)$$

Since n is odd and divides $2^n-2=2(2^{n-1}-1)$, we see that $n \mid (2^{n-1}-1)$. Therefore $2^{n-1}-1=kn$, i.e. $2^n-2=2kn$. Finally,

$$2^{m-1}=2^{2^n-2}=2^{2kn}$$

and

$$2^{m-1}-1=(2^n)^{2k}-1=(2^n-1)\left[2^{n(2k-1)}+\ldots+1\right]$$

Then:

$$\underbrace{2^n-1}_{m} \mid (2^{m-1}-1) \Rightarrow m \mid 2^m-2$$

EXAMPLE 3.46. If $\tau(n)$ is the number of all divisors of n and $\delta(n)$ is the product of all divisors of n, then:

$$\delta(n)=\sqrt{n^{\tau(n)}}$$

SOLUTION: Let the set of all divisors of n be $\{d_1,d_2,\ldots,d_{\tau(n)}\}$. This set can also be written as $\{\frac{n}{d_1},\frac{n}{d_2},\ldots,\frac{n}{d_{\tau(n)}}\}$. Therefore:

$$\delta(n)=d_1d_2\ldots d_{\tau(n)}=\frac{n^{\tau(n)}}{d_1d_2\ldots d_{\tau(n)}}$$

Then:

$$\delta^2(n)=n^{\tau(n)}$$

EXAMPLE 3.47. Prove

$$\left(\sum_{d|n}\tau(d)\right)^2=\sum_{d|n}\tau^3(d)$$

SOLUTION: From the properties of multiplicative functions and Theorem 3.12, both sides of the equation are multiplicative functions; therefore it suffices to consider the special case $n=p^\alpha$ when this identity reduces to the interesting identity from Examples 2.52 and A.3.

EXAMPLE 3.48 (MERSENNE PRIMES). If $2^m - 1$ is a prime, then m must be a prime too. Prove.

SOLUTION: If we assume m to be composite, i.e. $m = ab$, with $a, b > 1$, then:

$$2^m - 1 = 2^{ab} - 1 = (2^a)^b - 1^b = (2^a - 1)\left[2^{a(b-1)} + \ldots + 1\right]$$

i.e. $2^m - 1$ is also composite.

NOTES: This is just a one-way implication because if m is prime, $2^m - 1$ may be composite (for example, $2^{11} - 1 = 2047 = 23 \cdot 89$).

For prime numbers p, numbers $M_p = 2^p - 1$ are called *Mersenne numbers*. Some of these are primes (e.g., for $p = 2, 3, 5, 7, 9$), while some are composite (e.g., $p = 11$). Due to their many interesting properties, it is much easier to investigate the primality of Mersenne numbers M_p then of other integers of comparable size. Hence it is not surprising that some of the largest known primes are Mersenne primes. For example in 1968 the discovery of the then largest prime $2^{11213} - 1$ at the University of Illinois at Urbana–Champaign was celebrated by an appropriate postal stamp. Since then many more primes have been found among the Mersenne numbers, for example $2^{21701} - 1$, found in 1978 by two high-school students in California.

EXAMPLE 3.49 (PERFECT NUMBERS). Many civilizations found the properties of numbers not only interesting but also mystical. For example Ancient Greek mathematicians defined and investigated properties of *perfect numbers*, integers that equal the sum of their *proper* divisors. The first such number is 6, because:

$$6 = 1 + 2 + 3$$

The only perfect numbers < 10000 are $6, 28, 496$, and 8128. It is a nice programming exercise to determine other perfect numbers < 1000000.

Euclid knew that if $2^n - 1$ is a prime (then of course n must be prime, and $2^n - 1$ is a Mersenne prime), then:

$$2^{n-1}(2^n - 1)$$

is perfect.

Euler showed that an even number is perfect if and only if it takes the form

$$2^{p-1} M_p$$

where M_p is a Mersenne prime (not just any Mersenne number). To this day it is not known whether or not there are odd perfect numbers.

EXAMPLE 3.50. Show that if n is a perfect number, then:

$$\sum_{d|n} \frac{1}{d} = 2$$

SOLUTION: The equivalent definition of perfect numbers is that the sum of *all* of their divisors is twice the number itself ($\sigma(n) = 2n$). For example $1+2+3+6 = 2 \cdot 6$. With that in mind, consider

$$\sum_{d|n} \frac{1}{d} = \sum_{d|n} \frac{1}{n/d} = \frac{1}{n} \sum_{d|n} d = \frac{\sigma(n)}{n} = 2$$

4

Geometry

In Chapter 4 we present the simplest theorems and formulas about plane triangles. A few historical remarks should encourage the reader to think about the relation between mathematics and other exact sciences. We do not start from the axioms because we want to obtain the results more rapidly and in a more relaxed way.

4.1. Properties of Triangles

THEOREM 4.1 (PYTHAGOREAN THEOREM). *A square drawn on the hypotenuse of a right triangle is equal in area to the sum of the squares drawn on its sides.*

PROOF: In Euclid's *Elements*, the Pythagorean theorem is labeled I-47 (the forty-seventh theorem in the first book). The proof given here is not from the *Elements*; it is probably of Indian origin. (See also Examples 4.32 and 4.33.)

From the diagram in Fig. 4.1 we find

$$c^2 = (a+b)^2 - 4\left(\frac{1}{2}ab\right) = a^2 + b^2$$

The same conclusion can be reached without algebra if we compare the two diagrams.

COROLLARY: An immediate consequence of the Pythagorean theorem is the trigonometric identity $\sin^2\alpha + \cos^2\alpha = 1$. Indeed:

$$\sin^2\alpha + \cos^2\alpha = \left(\frac{a}{c}\right)^2 + \left(\frac{b}{c}\right)^2 = \frac{a^2+b^2}{c^2} = 1 \qquad \square$$

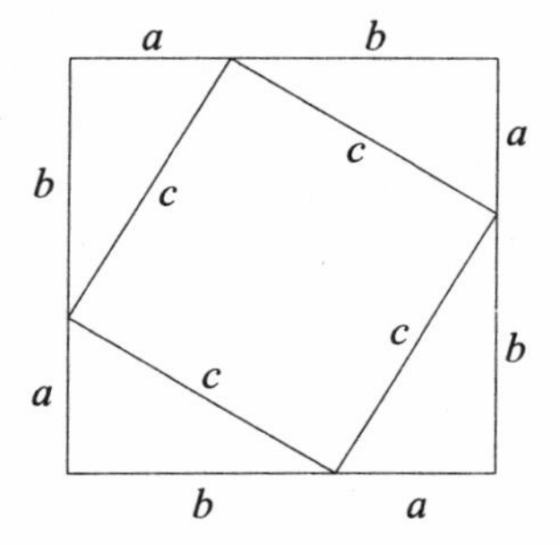

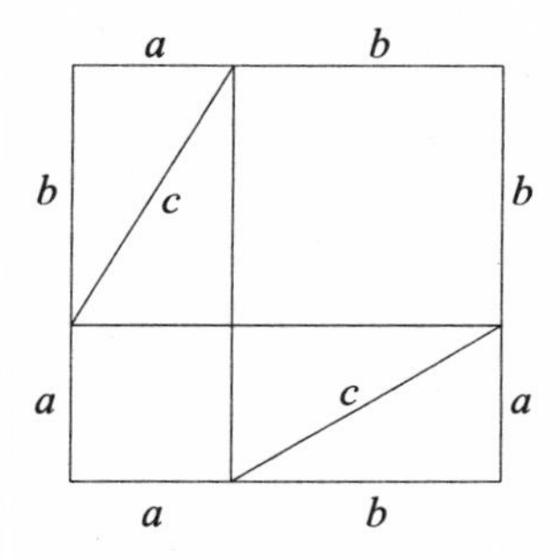

FIGURE 4.1. Proof of the Pythagorean theorem.

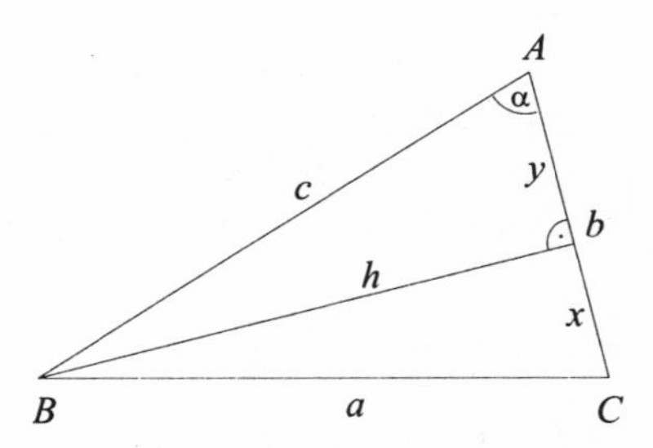

FIGURE 4.2. Proof of the law of cosines.

Among historians of mathematics there are different opinions about who discovered the Pythagorean theorem, but it is certain that it was known to Babylonian mathematicians a whole millennium before Pythagoras. We know that from the clay tablet, now called *Plimpton* 322, written in Babylon between 1900 and 1600 B.C., kept today in the Plimpton Library at Columbia University in New York.

We now use the Pythagorean theorem to prove the law of cosines.

THEOREM 4.2 (LAW OF COSINES). *In an arbitrary triangle, with notation as in Fig. 4.2,*

$$a^2 = b^2 + c^2 - 2bc\cos\alpha$$

PROOF: From the Pythagorean theorem we have

$$a^2 = x^2 + h^2 = (b - y)^2 + h^2$$

Since $y = c\cos\alpha$ and $h = c\sin\alpha$, we have Theorem 4.2. □

NOTE: The preceding derivation is not complete because we considered only the case when α is an acute angle. It is left to the reader to consider the case when α is obtuse.

Before proceeding with the proof of the law of sines, let us consider a few applications of the law of cosines.

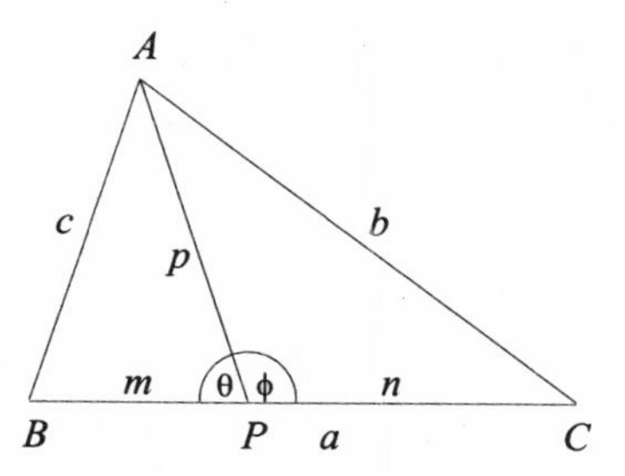

FIGURE 4.3. Proof of Stewart's theorem.

EXAMPLE 4.1 (STEWART'S THEOREM). With the notation as in Fig. 4.3, for an arbitrary segment $p = [AP], P \in [BC]$:

$$a(p^2 + mn) = mb^2 + nc^2$$

This theorem was derived by Stewart in 1746, but the first known proof was given by Simson in 1751. It is also quite probable that Archimedes knew it.

PROOF: Since $\theta = 180^\circ - \phi$, we have $\cos\theta = -\cos\phi$. Applying the law of cosines to the angle θ of the triangle $\triangle ABP$ leads to:

$$\cos\theta = \frac{p^2 + m^2 - c^2}{2pm}$$

Similarly in $\triangle ACP$:

$$\cos\phi = \frac{p^2 + n^2 - b^2}{2pn}$$

Since $m + n = a$ and $\cos\theta = -\cos\phi$, we find

$$a(p^2 + mn) = mb^2 + nc^2$$

EXAMPLE 4.2 (HERO'S FORMULA). Denote by h_a the altitude corresponding to the side a. Then the area of the triangle can be written as $P_\triangle = (1/2)ah_a$.

Since $h_a = c\sin\beta$, using the law of cosines for the angle β, we find

$$\begin{aligned}P_{\triangle} &= \frac{1}{2}ac\sin\beta \\ &= \frac{1}{2}ac\sqrt{1-\cos^2\beta} \\ &= \frac{1}{2}ac\sqrt{1-\frac{(a^2+c^2-b^2)^2}{4a^2c^2}} \\ &= \frac{1}{4}\sqrt{(a+b+c)(-a+b+c)(a-b+c)(a+b-c)} \\ &= \sqrt{s(s-a)(s-b)(s-c)}\end{aligned}$$

where s is the semiperimeter of the triangle:

$$s = \frac{a+b+c}{2}$$

This is Hero's famous formula. It is quite probable that Archimedes knew this formula, too. The original proof by Hero is given in Example 4.42.

EXAMPLE 4.3 (ANGLE OVER DIAMETER). Consider $\triangle XYZ$ inscribed in the circle K with center O and radius R such that the side $z = [XY]$ is a diameter of K.

Since $[OX] = [OY] = [OZ] = R$ and $z = 2R$, according to Stewart's theorem:

$$2R(R^2+R^2) = R(x^2+y^2)$$

that is,

$$x^2+y^2 = (2R)^2 = z^2$$

on the other hand from the law of cosines:

$$z^2 = x^2+y^2-2xy\cos\angle Z$$

Hence we find that $\cos\angle Z = 0$, i.e., the peripheral angle over a diameter, is a right angle.

EXAMPLE 4.4 (CENTRAL AND PERIPHERAL ANGLES I). Consider the special case of a central and a peripheral angle over a chord when one of the lines defining the peripheral angle is a diameter (Fig. 4.4).

According to Example 4.3 $\angle AXY$ is a right angle, so $\angle XYA = 90^\circ - \alpha$. Since $\triangle OXY$ is isosceles, we have $\angle YXO = \angle XYA = 90^\circ - \alpha$; therefore $\beta = 180^\circ - \angle YXO - \angle XYA = 180^\circ - 2(90^\circ - \alpha) = 2\alpha$.

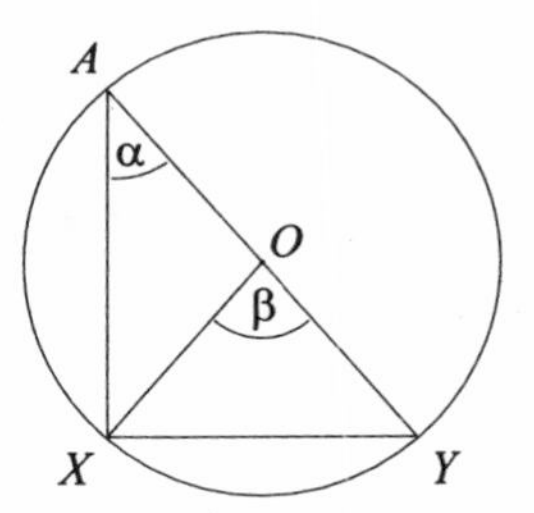

FIGURE 4.4. Central and peripheral angles I.

EXAMPLE 4.5 (CENTRAL AND PERIPHERAL ANGLES II). In general when the lines forming the central and peripheral angles do not coincide (Fig. 4.5), the angles α and β can be divided by the diameter containing the vertex A, so that the result from Example 4.4 can be used

$$\beta = \beta_1 + \beta_2 = 2\alpha_1 + 2\alpha_2 = 2(\alpha_1 + \alpha_2) = 2\alpha$$

Therefore an arbitrary peripheral angle is half of the central angle over the same chord. This also implies that all peripheral angles over the same chord are equal.

NOTE: It is left to the reader to complete the previous derivation by considering peripheral angles not having center O at their interior. The proof is similar, but instead of a sum, the difference of angles should be used. It is also left to the reader to show that the sum of peripheral angles on opposite sides of the same chord equals 180°. □

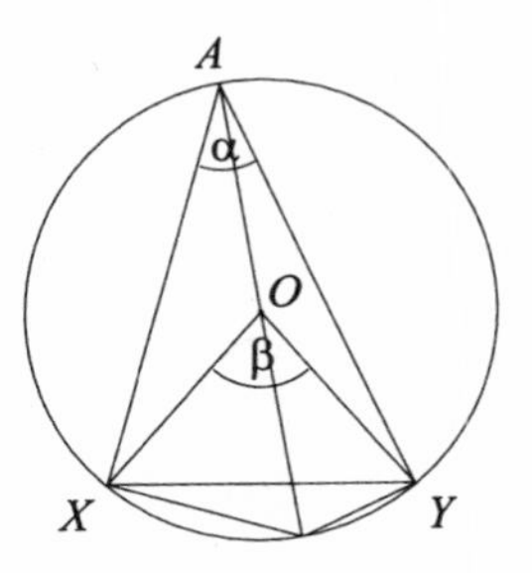

FIGURE 4.5. Central and peripheral angles II.

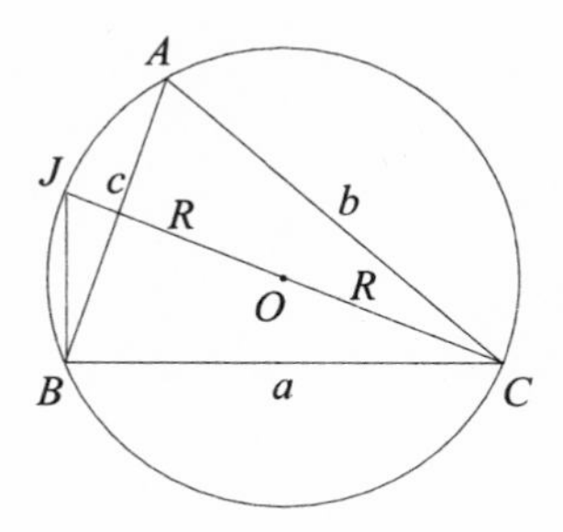

FIGURE 4.6. Proof of the law of sines.

Let us look at Fig. 4.6. The circle with center O and radius R is circumscribed about $\triangle ABC$. The point J is opposite to C with respect to O; hence $[JC]$ is a diameter, and $\angle JBC$ is a right angle. Since $\angle A$ and $\angle J$ are peripheral angles over the same chord, $\angle A = \angle J$ and $\sin \angle A = \sin \angle J$.

Therefore

$$\sin\alpha = \sin\angle A = \sin\angle J = \frac{[BC]}{[JC]} = \frac{a}{2R}$$

Similarly we find $\sin\beta = \frac{b}{2R}$, as well as $\sin\gamma = \frac{c}{2R}$, so finally we have Theorem 4.3.

THEOREM 4.3 (LAW OF SINES). *In an arbitrary triangle:*

$$\frac{a}{\sin\alpha} = \frac{b}{\sin\beta} = \frac{c}{\sin\gamma} = 2R$$

where R is the radius of the circumcircle of the triangle.

NOTE: Strictly speaking this derivation is not complete because we did not consider the case when α is an obtuse angle, but the omitted details are quite uninteresting.

EXAMPLE 4.6 (ABOUT ANGLE BISECTOR). Let the bisector of the angle α intersect $[BC]$ at L (Fig. 4.7). The law of sines applied to $\triangle ABL$ yields

$$\frac{m}{\sin(\alpha/2)} = \frac{c}{\sin\theta}$$

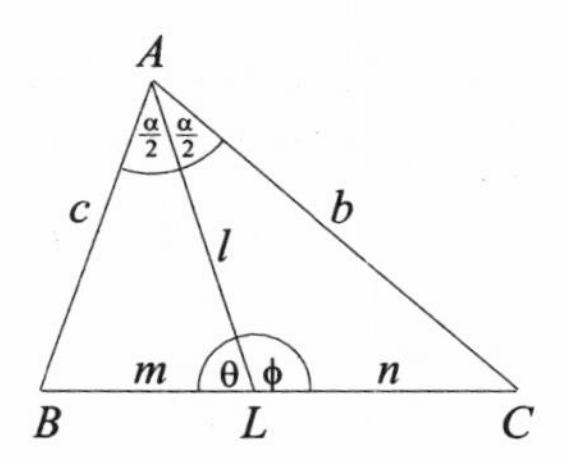

FIGURE 4.7. The angle bisector.

Similarly in $\triangle ALC$ we find

$$\frac{n}{\sin(\alpha/2)} = \frac{b}{\sin\phi}$$

Since $\sin\theta = \sin\phi$, we have

$$\frac{m}{c} = \frac{\sin(\alpha/2)}{\sin\theta} = \frac{\sin(\alpha/2)}{\sin\phi} = \frac{n}{b}$$

That is:

$$\frac{m}{n} = \frac{c}{b}$$

EXAMPLE 4.7 (LENGTH OF ANGLE BISECTOR). We now use Example 4.6 and Stewart's theorem to determine the length of the angle bisector $l_\alpha = [AL]$ (see Fig. 4.7).

From $m+n = a$ and $m/n = c/b$

$$m = \frac{ac}{b+c} \qquad n = \frac{ab}{b+c}$$

Hence from Stewart's theorem:

$$a\left[l_\alpha^2 + \frac{a^2bc}{(b+c)^2}\right] = \frac{ac}{b+c}b^2 + \frac{ab}{b+c}c^2$$

This implies

$$l_\alpha = \sqrt{bc\left[1 - \left(\frac{a}{b+c}\right)^2\right]}$$

EXAMPLE 4.8 (LENGTH OF MEDIAN). From the definition of the median, we have $m = n = a/2$, thus, according to Stewart's theorem:

$$a\left(t_a^2 + \frac{a^2}{4}\right) = \frac{a}{2}(b^2 + c^2)$$

That is:

$$t_a = \frac{1}{2}\sqrt{2(b^2 + c^2) - a^2}$$

EXAMPLE 4.9 (LENGTH OF ALTITUDE). The length of the altitude h_a can be found using Stewart's theorem, but the derivation from Hero's formula is much simpler:

$$\begin{aligned} h_a &= \frac{2}{a} P_\triangle \\ &= \frac{2}{a}\sqrt{s(s-a)(s-b)(s-c)} \end{aligned}$$

EXAMPLE 4.10 (CIRCUMCENTER). The circumcenter of a triangle, i.e., the center of the circumcircle (the circle circumscribed about the triangle), is found as the intersection of the perpendicular bisectors of the sides (see Fig. 4.8). Why? Because the circumcircle contains all three vertices of the triangle. This means that the circumcenter O is at the same distance from all three vertices. Hence:

$$O \in \text{perp bis}\,[AB] \qquad O \in \text{perp bis}\,[AC] \qquad O \in \text{perp bis}\,[BC]$$

That is:

$$O = \text{perp bis}\,[AB] \cap \text{perp bis}\,[AC] \cap \text{perp bis}\,[BC]$$

Of course when constructing the circumcenter O, it suffices to construct only the perpendicular bisectors of any two sides.

EXAMPLE 4.11 (INCENTER). The incenter of a triangle, i.e., the center of the incircle (the circle inscribed in the triangle), is found as the intersection of the angle bisectors (see Fig. 4.9). The incircle touches all three sides of the triangle. For a circle to touch both arms of the angle α, its center must lie on

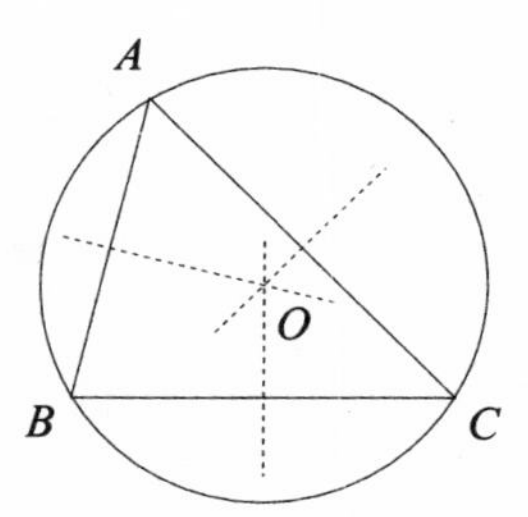

FIGURE 4.8. The circumcenter O of a triangle is at the intersection of the perpendicular bisectors of the sides.

the bisector of that angle. Similarly we find that it must lie on the bisectors of angles β and γ. Hence for the incenter I, we can write

$$I \in \text{bis}\,\alpha \qquad I \in \text{bis}\,\beta \qquad I \in \text{bis}\,\gamma$$

That is:

$$I = \text{bis}\,\alpha \cap \text{bis}\,\beta \cap \text{bis}\,\gamma$$

Again when constructing I, any two bisectors are sufficient.

EXAMPLE 4.12 (AREA OF TRIANGLE AND CIRCUMRADIUS R). Since $\sin\beta = \frac{b}{2R}$, we have

$$P_\triangle = \frac{1}{2}ah_a = \frac{1}{2}ac\sin\beta = \frac{abc}{4R}$$

where R is the circumradius.

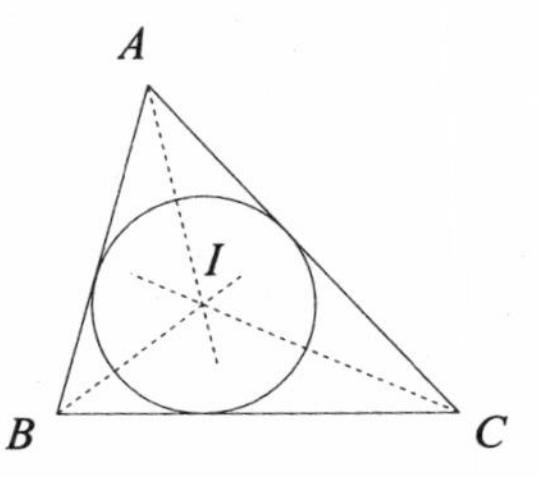

FIGURE 4.9. The incenter I of a triangle is at the intersection of the angle bisectors.

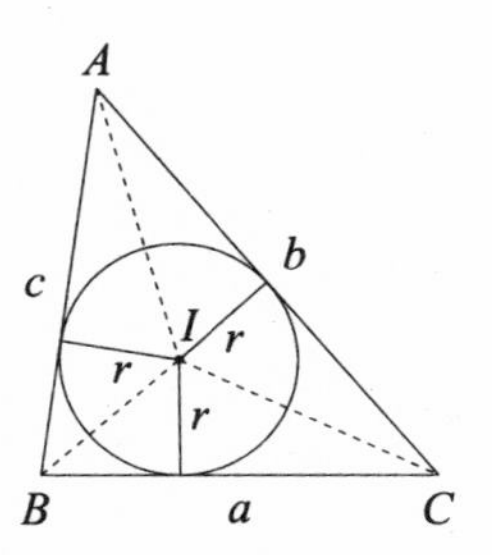

FIGURE 4.10. Proof of $P_{\triangle} = sr$.

EXAMPLE 4.13 (CIRCUMRADIUS). From Example 4.12 and Hero's formula:

$$R = \frac{abc}{4\sqrt{s(s-a)(s-b)(s-c)}}$$

EXAMPLE 4.14 (AREA OF TRIANGLE AND INRADIUS r). If r is the inradius (see Fig. 4.10) we see that:

$$\begin{aligned} P_{ABC} &= P_{ABI} + P_{BCI} + P_{CAI} \\ &= \frac{ar}{2} + \frac{br}{2} + \frac{cr}{2} \\ &= sr \end{aligned}$$

EXAMPLE 4.15 (INRADIUS). From Example 4.14 and Hero's formula, we have:

$$r = \sqrt{\frac{(s-a)(s-b)(s-c)}{s}}$$

EXAMPLE 4.16 (EULER'S FORMULA). From Examples 4.13 and 4.14 we can derive a formula due to Euler that relates the sides of a triangle to its inradius, circumradius, and semiperimeter:

$$4rRs = abc$$

This relation was probably known to geometers in Ancient Greece.

THEOREM 4.4 (CEVA'S THEOREM). *If the cevians $[AX]$, $[BY]$, and $[CZ]$ of the triangle $\triangle ABC$ are concurrent, then:*

$$\frac{[BX][CY][AZ]}{[CX][AY][BZ]} = 1$$

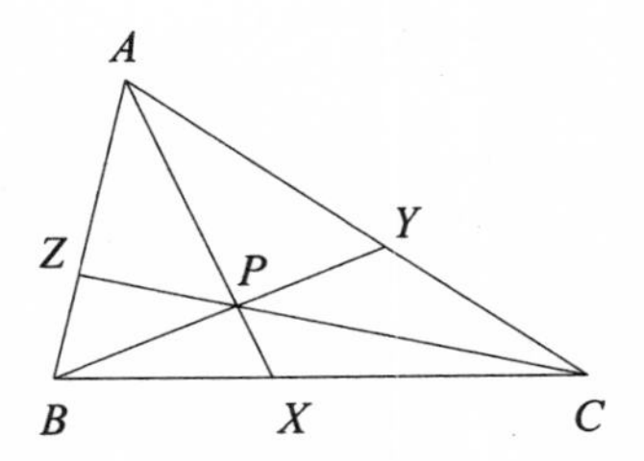

FIGURE 4.11. Proof of Ceva's theorem.

PROOF: We now prove Ceva's theorem from 1678 (see Fig. 4.11). But first let us define two useful terms: The line segment defined by the vertex of a triangle and a point on the opposite side is called a *cevian*; three lines are said to be *concurrent* if all three pass through one point.

In Example 4.10 we found that the perpendicular bisectors of the sides of a triangle are concurrent. (Their intersection is the circumcenter of the triangle.) Similarly in Example 4.11 we showed that the angle bisectors of a triangle are concurrent (with the incenter as their intersection). Note: Angle bisectors are cevians, while the perpendicular side bisectors are not.

Consider Fig. 4.11. Let cevians $[AX]$, $[BY]$, and $[CZ]$ be concurrent. Then:

$$\frac{[BX]}{[CX]} = \frac{P_{ABX}}{P_{ACX}} = \frac{P_{PBX}}{P_{PCX}}$$

Since $a/b = x/y$ implies $a/b = x/y = (a-x)/(b-y)$, we have

$$\frac{[BX]}{[CX]} = \frac{P_{ABX} - P_{PBX}}{P_{ACX} - P_{PCX}} = \frac{P_{ABP}}{P_{ACP}}$$

We find similar formulas for $[CY]/[AY]$ and $[AZ]/[BZ]$; therefore:

$$\frac{[BX][CY][AZ]}{[CX][AY][BZ]} = \frac{P_{ABP}P_{BCP}P_{ACP}}{P_{ACP}P_{ABP}P_{BCP}} = 1 \qquad \square$$

The converse theorem (Theorem 4.5) is also easy to prove (see Example 4.36):

THEOREM 4.5. *If cevians $[AX]$, $[BY]$, and $[CZ]$ of $\triangle ABC$ divide sides $[BC]$, $[AC]$, and $[AB]$ so that:*

$$\frac{[BX][CY][AZ]}{[CX][AY][BZ]} = 1$$

then they are concurrent.

EXAMPLE 4.17 (ANGLE BISECTORS ARE CONCURRENT). For the angle bisectors, we have

$$\frac{[BX][CY][AZ]}{[CX][AY][BZ]} = \frac{c}{b}\frac{a}{c}\frac{b}{a} = 1$$

Therefore according to Theorem 4.5, they are concurrent. A different proof was given in Example 4.11, where we saw that their intersection is at the *incenter* of the triangle.

EXAMPLE 4.18 (MEDIANS ARE CONCURRENT). The medians of a triangle divide its sides so that $[BX] = [CX]$, $[CY] = [AY]$, and $[AZ] = [BZ]$; therefore:

$$\frac{[BX][CY][AZ]}{[CX][AY][BZ]} = 1$$

That is, the medians are concurrent. Their intersection is called the *centroid* of the triangle.

EXAMPLE 4.19 (ALTITUDES ARE CONCURRENT). For the altitudes of a triangle, we have

$$\frac{[BX][CY][AZ]}{[CX][AY][BZ]} = \frac{c\cos\beta}{b\cos\gamma}\frac{a\cos\gamma}{c\cos\alpha}\frac{b\cos\alpha}{a\cos\beta} = 1$$

This proves that they are concurrent. Their intersection is called the *orthocenter* of the triangle. □

If points $A_1,\ldots,A_n$ have masses $m_1,\ldots,m_n$, we have a *system of material points*, usually denoted by $\{(m_1,A_1),\ldots,(m_n,A_n)\}$.

A point T is the *centroid* of the system $\{(m_1,A_1),\ldots,(m_n,A_n)\}$ of material points if the following is satisfied:

$$m_1\overrightarrow{A_1T} + \ldots + m_n\overrightarrow{A_nT} = 0 \tag{4.1}$$

THEOREM 4.6. *For an arbitrary system of material points*

$$\{(m_1,A_1),\ldots,(m_n,A_n)\}$$

there is one and only one centroid.

PROOF: Let the origin of the coordinate system be denoted by O. If both sides of the defining equality (4.1) are augmented by $m_1\overrightarrow{OA_1}+\ldots+m_n\overrightarrow{OA_n}$, then:

$$(m_1+\ldots+m_n)\overrightarrow{OT}=m_1\overrightarrow{OA_1}+\ldots+m_n\overrightarrow{OA_n}$$

This implies

$$\overrightarrow{OT}=\frac{m_1\overrightarrow{OA_1}+\ldots+m_n\overrightarrow{OA_n}}{m_1+\ldots+m_n}$$

NOTE: The reader is probably familiar with the fact that if T is the centroid of the system $\{(m_1,A_1),\ldots,(m_n,A_n)\}$ and if we add another material point to it, e.g., (m_{n+1},A_{n+1}), the centroid of the new system is the centroid of $\{(m,T),(m_{n+1},A_{n+1})\}$, where $m=m_1+\ldots+m_n$.

Indeed:

$$\overrightarrow{OT'}=\frac{m_1\overrightarrow{OA_1}+\ldots+m_n\overrightarrow{OA_n}+m_{n+1}\overrightarrow{OA_{n+1}}}{m_1+\ldots+m_n+m_{n+1}}=\frac{m\overrightarrow{OT}+m_{n+1}\overrightarrow{OA_{n+1}}}{m+m_{n+1}}$$

We see that T' lies on the segment $[A_{n+1}T]$ and further:

$$[TT']=\frac{m_{n+1}}{m+m_{n+1}}[TA_{n+1}] \tag{4.2}$$

□

THEOREM 4.7 (ARCHIMEDES' THEOREM). *The medians of a triangle intersect at the centroid of a triangle, and they divide each other in* $1:2$ *ratio.*

PROOF: From Theorem 4.6, independent of Ceva's theorem, we find that the medians are concurrent and that they divide each other with the ratio $1:2$. The centroid T of the triangle $\triangle ABC$ is the centroid of the system:

$$\{(m,A),(m,B),(m,C)\}$$

Hence as mentioned earlier, T is also the centroid of the following systems:

$$\{(2m,T_{AB}),(m,C)\} \qquad \{(2m,T_{AC}),(m,B)\} \qquad \{(2m,T_{BC}),(m,A)\}$$

where T_{AB}, T_{AC}, and T_{BC} are centroids of the segments $[AB]$, $[AC]$, and $[BC]$, respectively.

Hence $T \in [CT_{AB}]$, $T \in [BT_{AC}]$, and $T \in [AT_{BC}]$. This implies that the medians are concurrent, since their intersection is the centroid T.

From Eq. (4.2) we find that the medians divide each other in a $1:2$ ratio. This proves Theorem 4.7. □

Earlier we used Stewart's theorem to find the length of the median t_a:

$$t_a = \frac{1}{2}\sqrt{2(b^2+c^2)-a^2}$$

We use this result in Examples 4.20 and 4.21.

EXAMPLE 4.20 (TWO SIMILAR TRIANGLES). Consider $\triangle ABC$ and use its medians to form a new triangle $\triangle A_1B_1C_1$. If $\triangle A_2B_2C_2$ is constructed from the medians of $\triangle A_1B_1C_1$, then triangles $\triangle ABC$ and $\triangle A_2B_2C_2$ are similar.

The squares of the sides of $\triangle A_1B_1C_1$ are given by:

$$a_1^2 = t_a^2 = \frac{1}{4}[2(b^2+c^2)-a^2]$$
$$b_1^2 = t_b^2 = \frac{1}{4}[2(a^2+c^2)-b^2]$$
$$c_1^2 = t_c^2 = \frac{1}{4}[2(a^2+b^2)-c^2]$$

The squares of the sides of $\triangle A_2B_2C_2$ are given by:

$$a_2^2 = t_{a_1}^2 = \frac{9}{16}a^2 \qquad b_2^2 = t_{b_1}^2 = \frac{9}{16}b^2 \qquad c_2^2 = t_{c_1}^2 = \frac{9}{16}c^2$$

Hence we find that $\triangle ABC$ and $\triangle A_2B_2C_2$ are similar, with a factor of $3/4$.

EXAMPLE 4.21 (LEIBNIZ'S THEOREM). Let T be the centroid of $\triangle ABC$. Then for an arbitrary point M, we have Leibniz's theorem (see Fig. 4.12):

$$[MA]^2+[MB]^2+[MC]^2 = 3[MT]^2+[AT]^2+[BT]^2+[CT]^2$$

Let T_1 be the center of $[BC]$. Then $[MT_1]$ is a median for $\triangle MBC$; from Example 4.8 we know that:

$$[MT_1]^2 = \frac{1}{4}(2[MB]^2+2[MC]^2-[BC]^2)$$

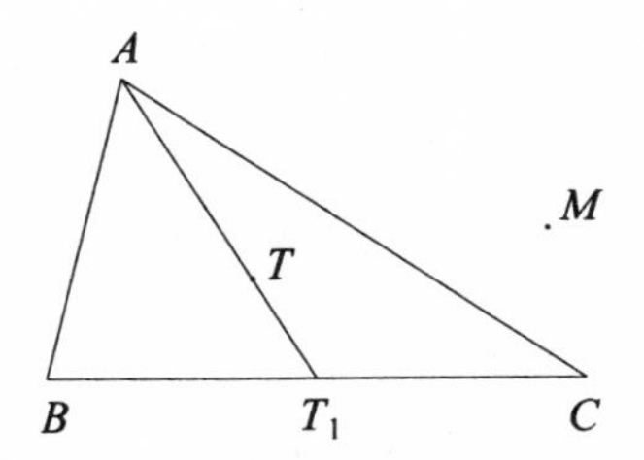

FIGURE 4.12. Proof of Leibniz's theorem.

Similarly $[AT_1]$ is a median of $\triangle ABC$ and:

$$[AT_1]^2 = \frac{1}{4}(2[AB]^2 + 2[AC]^2 - [BC]^2)$$

According to Archimedes' theorem $[AT] = \frac{2}{3}[AT_1]$ and $[TT_1] = \frac{1}{3}[AT_1]$. Now using Stewart's theorem in $\triangle MAT_1$ we find

$$[MT]^2 = \frac{1}{3}\left\{[MA]^2 + [MB]^2 + [MC]^2 - \frac{1}{3}([AB]^2 + [AC]^2 + [BC]^2)\right\}$$

From the formula for the length of the median we obtain

$$[AB]^2 + [AC]^2 + [BC]^2 = \frac{4}{3}(t_a^2 + t_b^2 + t_c^2) = 3([AT]^2 + [BT]^2 + [CT]^2)$$

This is an interesting identity in its own right. Finally:

$$[MA]^2 + [MB]^2 + [MC]^2 = 3[MT]^2 + [AT]^2 + [BT]^2 + [CT]^2 \qquad \square$$

It is often the case in mathematics that a new concept results in easier proofs of already known theorems and also lets us derive new results. Such is the case with the *moment of inertia*, too.

The *moment of inertia* of the system $\{(m_1, A_1), \ldots, (m_n, A_n)\}$ with respect to the arbitrary point M is defined as:

$$J_M = m_1[A_1M]^2 + \ldots + m_n[A_nM]^2$$

In Theorems 4.8 and 4.9 we prove two important properties of the moment of inertia.

THEOREM 4.8 (LAW OF STEINER AND LAGRANGE). *For the moment of inertia with respect to the arbitrary point M, we have*

$$J_M = J_T + m[MT]^2$$

where $m = m_1 + \ldots + m_n$, while T is the centroid of the system.

PROOF: The following sequence of equalities leads us to the desired formula:

$$\begin{aligned} J_M &= m_1[A_1M]^2 + \ldots + m_n[A_nM]^2 \\ &= m_1(\overrightarrow{A_1M})^2 + \ldots + m_n(\overrightarrow{A_nM})^2 \\ &= m_1(\overrightarrow{A_1T} + \overrightarrow{TM})^2 + \ldots + m_n(\overrightarrow{A_nT} + \overrightarrow{TM})^2 \\ &= J_T + 2(m_1\overrightarrow{A_1T} + \ldots + m_n\overrightarrow{A_nT})\overrightarrow{TM} + m[TM]^2 \\ &= J_T + m[MT]^2 \end{aligned}$$

□

COROLLARY: An immediate consequence of the law of Steiner and Lagrange is that the moment of inertia is the least when taken with respect to the centroid. Indeed since $m[MT]^2 \geq 0$, the following inequality holds

$$J_M \geq J_T$$

with the equality if and only if $M = T$.

THEOREM 4.9 (JACOBI'S THEOREM). *The moment of inertia with respect to the centroid of a system is given by:*

$$J_T = \frac{1}{m} \sum_{1 \leq i \leq j \leq n} m_i m_j r_{ij}^2$$

where $r_{ij} = [A_iA_j]$ and $m = m_1 + \ldots + m_n$.

PROOF: Apply the law of Steiner and Lagrange to the special case when $M = A_k$, where A_k is a point of the system, to obtain

$$J_{A_k} = J_T + m[A_kT]^2$$

Multiply both sides of this equality by m_k, then sum both sides over k to obtain

$$J_T = \frac{1}{m} \sum_{1 \leq i \leq j \leq n} m_i m_j r_{ij}^2$$

□

EXAMPLE 4.22 (MOMENT OF INERTIA OF TRIANGLE). Let us calculate the moment of inertia of $\triangle ABC$ with respect to its centroid T. If we substitute the unit masses for the vertices of the triangle, by definition we have

$$\begin{aligned} J_T &= [AT]^2 + [BT]^2 + [CT]^2 \\ &= \left(\frac{2}{3}t_a\right)^2 + \left(\frac{2}{3}t_b\right)^2 + \left(\frac{2}{3}t_c\right)^2 \\ &= \frac{4}{9}(t_a^2 + t_b^2 + t_c^2) \end{aligned}$$

On the other hand from Jacobi's theorem, we have

$$J_T = \frac{1}{3}(a^2 + b^2 + c^2)$$

Thus we have another proof of the identity:

$$t_a^2 + t_b^2 + t_c^2 = \frac{3}{4}(a^2 + b^2 + c^2)$$

which we found earlier using the formula for the length of a median.

EXAMPLE 4.23 (LEIBNIZ'S THEOREM AGAIN). If we look at the moment of inertia of a triangle with respect to the arbitrary point M, from the law of Steiner and Lagrange and the previous example we find

$$\begin{aligned} J_M &= J_T + 3[MT]^2 \\ &= 3[MT]^2 + [AT]^2 + [BT]^2 + [CT]^2 \end{aligned}$$

By definition:

$$J_M = [MA]^2 + [MB]^2 + [MC]^2$$

In a very simple and elegant manner we derived Leibniz's theorem:

$$[MA]^2 + [MB]^2 + [MC]^2 = 3[MT]^2 + [AT]^2 + [BT]^2 + [CT]^2$$

EXAMPLE 4.24 (STEWART'S THEOREM AGAIN). Let us consider how to use the concept of the moment of inertia to derive Stewart's theorem. Instead of the unit or equal masses, we now need the masses proportional to segments n and m at points B and C, respectively. We do not need mass at the vertex A.

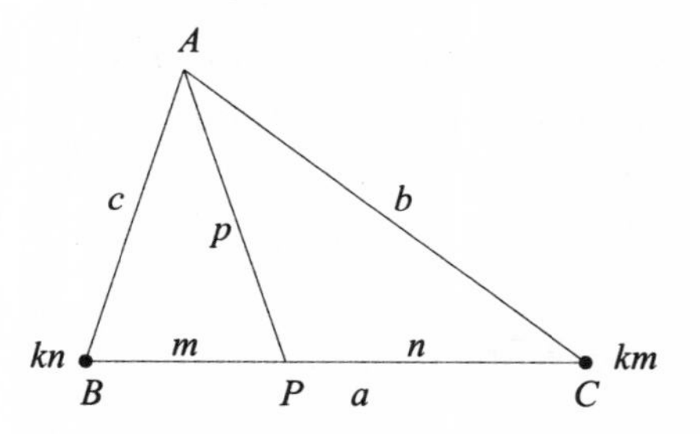

FIGURE 4.13. Another proof of Stewart's theorem.

Thus we consider the system $\{(kn,B),(km,C)\}$, where k is the proportionality constant. Such masses were chosen to have the centroid at P, the foot of the cevian p (see Fig. 4.13). According to the law of Steiner and Lagrange:

$$J_A = J_P + (kn+km)[AP]^2$$

Since:

$$J_A = kmb^2 + knc^2 \qquad J_P = knm^2 + kmn^2$$

We find

$$mb^2 + nc^2 = (m+n)p^2 + mn(m+n)$$

That is:

$$a(p^2+mn) = mb^2 + nc^2$$

EXAMPLE 4.25 (MASSES PROPORTIONAL TO SIDES). If we substitute $m_A = ka$ for A, $m_B = kb$ for B, and $m_C = kc$ for C, where k is some proportionality constant, the centroid of the system is at the incenter of $\triangle ABC$.

From Example 4.6 the centroid of the system $\{(kb,B),(kc,C)\}$ is at the foot of the bisector of α. Similarly centroids of the other two pairs of these three points are at feet of the corresponding angle bisectors. Therefore the centroid of such system is at its incenter.

EXAMPLE 4.26 (EULER'S LINE). Consider $\triangle ABC$ in Fig. 4.14. Let P, Q, and R be centers of sides $[BC]$, $[AC]$, and $[AB]$, respectively. The triangle $\triangle PQR$ is

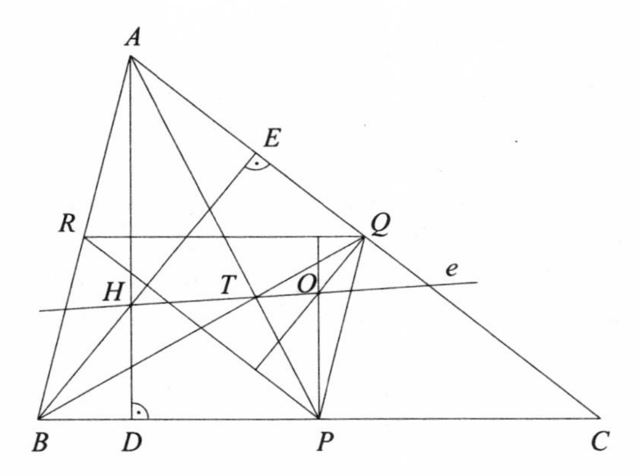

FIGURE 4.14. Euler's line.

the median triangle of $\triangle ABC$. These two triangles are similar, and it is obvious that the sides of $\triangle PQR$ are halves of the corresponding sides of $\triangle ABC$.

Let H be the *orthocenter* of $\triangle ABC$, T its centroid, and O its circumcenter.

Since the perpendicular bisectors of the sides of $\triangle ABC$ are altitudes of $\triangle PQR$, for $\triangle PQR$ the point O is the orthocenter. Since medians of $\triangle ABC$ bisect the sides of $\triangle PQR$, the centroids of these two triangles coincide. Therefore:

$$[PO] = \frac{1}{2}[AH] \qquad [PT] = \frac{1}{2}[AT]$$

In addition, since lines AH and PO are parallel and $T \in AP$, we find that:

$$\angle TAH = \angle TPO$$

From this we see that triangles $\triangle AHT$ and $\triangle POT$ are similar, hence:

$$\angle HTA = \angle OTP$$

Therefore points H, T, and O are collinear, T is between H and O, and

$$[HT] = 2[TO]$$

NOTE: The line containing the orthocenter, the centroid, and the circumcenter is called *Euler's line* of the triangle. Let us just mention the theorem by

Brianchon and Poncelet from 1820, which states that the center of $[OH]$ is the center of the *nine-point circle*, which contains feet of the altitudes, centers of the sides, and centers of the segments defined by the orthocenter and vertices of the triangle. The radius of the nine-point circle is $\frac{1}{2}R$. In 1822 Feuerbach showed that the nine-point circle touches the incircle and all three excircles* of the triangle. The nine-point circle is also called *Euler's circle* (in France), and *Feuerbach's circle* (in Germany). It can be proved that if R is the circumradius, r the inradius, and r_a, r_b, and r_c the exradii, the following is true:

$$4R = r_a + r_b + r_c - r$$

In addition:

$$\frac{1}{r_a} + \frac{1}{r_b} + \frac{1}{r_c} = \frac{1}{r}$$

EXAMPLE 4.27 (HAMILTON–SYLVESTER'S THEOREM). This theorem states that:

$$\overrightarrow{OH} = \overrightarrow{OA} + \overrightarrow{OB} + \overrightarrow{OC}$$

To prove this, write the definition of the centroid $\overrightarrow{AT} + \overrightarrow{BT} + \overrightarrow{CT} = 0$, then observe that:

$$\begin{aligned}\overrightarrow{OH} &= 3\overrightarrow{OT} \\ &= \overrightarrow{OA} + \overrightarrow{AT} + \overrightarrow{OB} + \overrightarrow{BT} + \overrightarrow{OC} + \overrightarrow{CT} \\ &= \overrightarrow{OA} + \overrightarrow{OB} + \overrightarrow{OC}\end{aligned}$$

EXAMPLE 4.28 (SEGMENT $[OT]$). In this example we derive the expression for the segment $[OT]$ in terms of the sides of the triangle only. If we substitute the unit masses in the vertices of the triangle, the law of Steiner and Lagrange applied to the circumcenter O yields

$$J_O = J_T + 3[OT]^2$$

From the definition of the moment of inertia and Example 4.22, we know that:

$$J_O = 3R^2 \qquad J_T = \frac{1}{3}(a^2 + b^2 + c^2)$$

*A circle touching one side of the triangle and the extensions of the other two sides is called the *excircle* or the *escribed circle* of the triangle. There are three excircles; their radii (the *exradii* of the triangle) are $r_a = P_\triangle/(s-a)$, $r_b = P_\triangle/(s-b)$, and $r_c = P_\triangle/(s-c)$.

Then:

$$[OT]^2 = R^2 - \frac{1}{9}(a^2 + b^2 + c^2)$$

NOTE: Earlier we found the expression for R in terms of a, b, and c; therefore the same can be done for $[OT]$ and segments $[OH]$ and $[TH]$, too.

EXAMPLE 4.29 (SEGMENT $[OI]$). Let us determine the expression for $[OI]$. As in Example 4.25 we substitute the masses ka, kb, and kc in the vertices A, B, and C, respectively, so that the centroid of the system is at the incenter I.

The law of Steiner and Lagrange gives

$$J_O = J_I + k(a+b+c)[OI]^2$$

Jacobi's theorem yields

$$J_I = \frac{k^2(abc^2 + ab^2c + a^2bc)}{k(a+b+c)} = kabc$$

From the definition of the moment of inertia it follows that:

$$J_O = k(a+b+c)R^2$$

Therefore:

$$[OI]^2 = \frac{J_O - J_I}{k(a+b+c)} = R^2 - \frac{abc}{a+b+c}$$

Observe that (see Examples 4.13 and 4.15):

$$\frac{abc}{a+b+c} = 2Rr$$

so finally:

$$[OI]^2 = R(R-2r)$$

This formula was first discovered by Euler. Since $[OI]^2 \geq 0$, it implies that $R \geq 2r$. In Example 4.47 we give another proof of this formula.

4.2. Analogies in Geometry

The moment of inertia was first defined by Euler in his work on the dynamics of rigid bodies. This new concept allowed Euler to simplify significantly the treatment of the whole field, but it was not the first useful contact of mechanics with geometry.

In the first of two books *On the Sphere and Cylinder*, Archimedes proves that if a cone is inscribed in a hemisphere that is inscribed in a cylinder, then the volumes of these three figures have the ratio $1:2:3$. According to legend, Archimedes was so proud of this discovery that he wanted a sphere inscribed in a cylinder with their ratio engraved on his tombstone. But although the geometric proof given by Archimedes is of the utmost elegance and perfection, the impression remains that Archimedes made this discovery in some other way. The following excerpt from his long-lost book* *The Method* proves that Archimedes really discovered the ratio using a method he called mechanical, and he followed the tradition of Greek geometers of merely proving theorems without indicating how they were discovered:

> certain things first became clear to me by a mechanical method, although they had to be demonstrated by geometry afterwards because their investigation by the said method did not furnish an actual demonstration.

Archimedes' mechanical method is essentially integration, a method discovered by mathematicians much later in the sixteenth century. The rigor that guarantees integration is not just a way of making discoveries but also of providing proofs was achieved still later, in the nineteenth century.

We can often use the analogy between optics and geometry. For example if we are given two points A and B on the same side of the line l (Fig. 4.15), and we wish to find the point $L \in l$ such that the sum $[AL]+[BL]$ is less than for any other point, we recall Fermat's principle from optics, which states that a ray of light travels between two points by the (optically) shortest path.

This gives us an idea† which we can use for a rigorous geometric proof. If B_1 is the point symmetric to B with respect to l (in optics B_1 is the mirror

*In 1906 Heiberg found a copy of *The Method* in a monastery in Constantinople.

†From the historic point of view, ideas went in the opposite direction: Hero knew that light is reflected so that the angle of incidence equals the angle of reflection. From that he proved the minimality of the sum $[AL]+[BL]$. Later Fermat, generalizing this and other similar examples, formulated the principle of minimality.

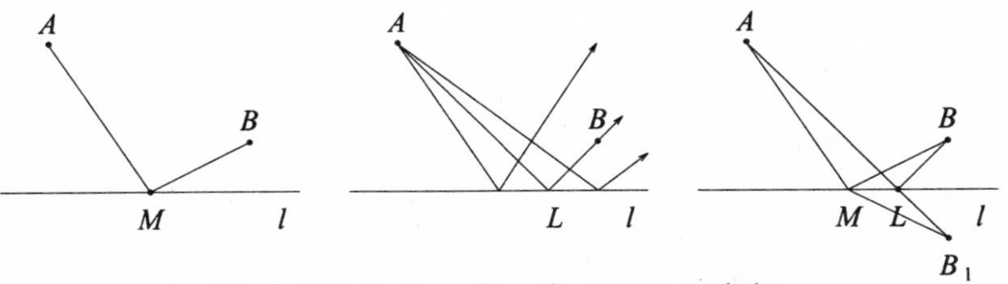

FIGURE 4.15. Analogy with optics suggests a solution.

image of B) and if M is an arbitrary point on l, we see that:

$$[AM]+[BM]=[AM]+[B_1M]\leq[AB_1]=[AL]+[B_1L]=[AL]+[BL]$$

Hence the optimal point L can be constructed as the intersection of lines l and AB_1.

Let us now consider another minimization problem first proposed by Fermat in a letter to Torricelli. The problem is to find the point F for which the sum of distances from the vertices of $\triangle ABC$ has a minimum, the so-called Fermat's point.

In this case, too, we can use the analogy with the optics, but the mechanical analogy is more elegant. Take three strings and attach weights X, Y, and Z of equal masses at their ends. Tie the free ends of the strings together, then arrange this device as in Fig. 4.16. If friction in the system is negligible, this mechanical system will reach equilibrium, the position of minimum energy. This equilibrium is characterized by the lowest possible position of the weights; hence the sum $[AX]+[BY]+[CZ]$ has a maximum. If the equilibrium position of the knot is at R, then since the sum $[RA]+[AX]+[RB]+[BY]+[RC]+[CZ]$ is constant, the sum $[RA]+[RB]+[RC]$ has a minimum, i.e., $R\equiv F$.

Instead of creating this simulator and measuring the coordinates of the equilibrium point $R\equiv F$, let us continue our thought experiment. Recall that a mechanical system is in equilibrium if the vector sum of all forces is zero. Since all weights are equal, the only position of the knot where all forces cancel each other is the one where vectors $\overrightarrow{FA}$, $\overrightarrow{FB}$, and $\overrightarrow{FC}$ form angles of $120°$.

We now proceed with the proof of these facts. If triangle $\triangle ABC$ and an arbitrary point M are rotated around C for $60°$ (let A_1, B_1, and M_1 be images of A, B, and M, respectively, $C_1\equiv C$), we see that $\triangle MM_1C$ is equilateral, hence $[MM_1]=[MC]$. Also $[M_1B_1]=[MB]$. Therefore:

$$[MA]+[MB]+[MC]=[AM]+[MM_1]+[M_1B_1]\geq[AB_1]$$

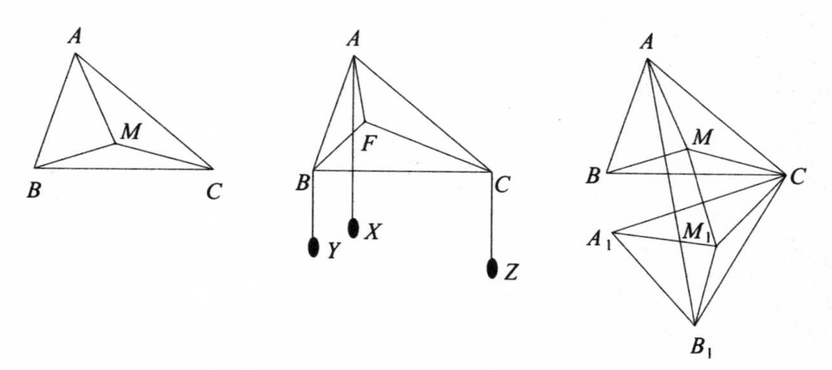

FIGURE 4.16. Fermat's point.

The sum $[MA]+[MB]+[MC]$ has a minimum for $M \equiv F$; hence F and F_1 must lie on $[AB_1]$. Then from F segment $[AC]$ is seen at 120°, because $\angle CFF_1 = 60°$. Similarly if the rotation is made about B, segment $[AB]$ is seen at 120° from F. Therefore the mechanical analogy gave us the right direction.

Probably the simplest way of constructing Fermat's point F of $\triangle ABC$ is to construct equilateral triangles $\triangle ABZ$ and $\triangle AYC$ over sides $[AB]$ and $[AC]$ in the exterior of $\triangle ABC$, and then to find F at the intersection of lines $[BY]$ and $[CZ]$. (See also Examples 4.38 and 4.45.)

4.3. Two Geometric Tricks

At the end of this chapter we examine two mathematical tricks first published in 1892 in the first edition of Ref. [49], which states that Euclid's *Elements* were followed by three collections of problems, unfortunately lost through the centuries. One of these three books contained mathematical tricks and puzzles, showing that Euclid thought such problems were useful: While we look for the error in a proof, a lot can be learned.

EXAMPLE 4.30 (OBTUSE AND RIGHT ANGLE ARE EQUAL!). At point A at right angles with $[AB]$, construct an arbitrary segment $[AC]$. At B at an angle of, e.g., 100°, construct $[BD]$ equal in length to $[AC]$. Let the perpendicular bisectors of $[AB]$ and $[CD]$ intersect at M (see Fig. 4.17).

Since M lies on the perpendicular bisector of $[AB]$, we see that $[MA] = [MB]$. Similarly $[MC] = [MD]$. Since also $[AC] = [BD]$, we find that the

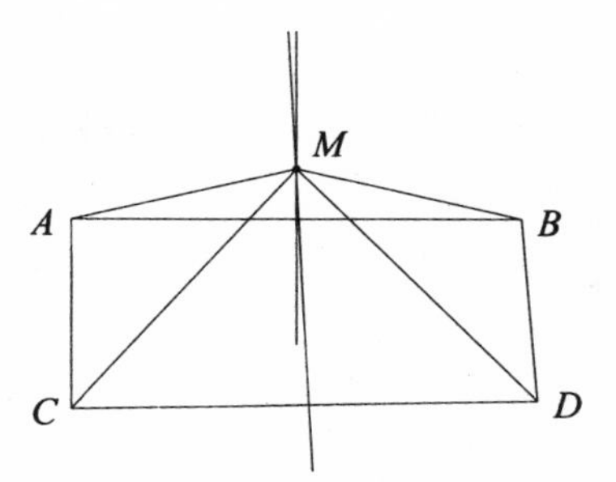

FIGURE 4.17. An obtuse and right angles are equal! Or perhaps not!

triangles $\triangle MAC$ and $\triangle MBD$ are congruent; hence:

$$\angle MAC \cong \angle MBD$$

In addition, since $\triangle ABM$ is isosceles:

$$\angle MAB \cong \angle MBA$$

Finally:

$$\begin{aligned} 90^\circ &= \angle BAC \\ &= \angle MAC - \angle MAB \\ &= \angle MBD - \angle MBA \\ &= \angle ABD \\ &= 100^\circ \end{aligned}$$

EXAMPLE 4.31 (ALL TRIANGLES ARE EQUILATERAL!). Consider the intersection of the bisector of $\angle A$ and the perpendicular bisector of $[BC]$ (see Fig. 4.18). If these two lines coincide, then $[AB] = [AC]$.

If they intersect at D, then consider the feet of perpendiculars from D at lines AB and AC, points E and F, respectively.

Triangles $\triangle ADE$ and $\triangle ADF$ are congruent because they have two pairs of congruent angles ($\angle DAE \cong \angle DAF$ and $\angle DEA \cong \angle DFA$) and one common side. Hence $[DE] \cong [DF]$; therefore triangles $\triangle BDE$ and $\triangle CDF$ are congruent because they have two pairs of congruent sides ($[BD] \cong [CD]$ and $[DE] \cong [DF]$) and a pair of congruent angles ($\angle DEB \cong \angle DFC$).

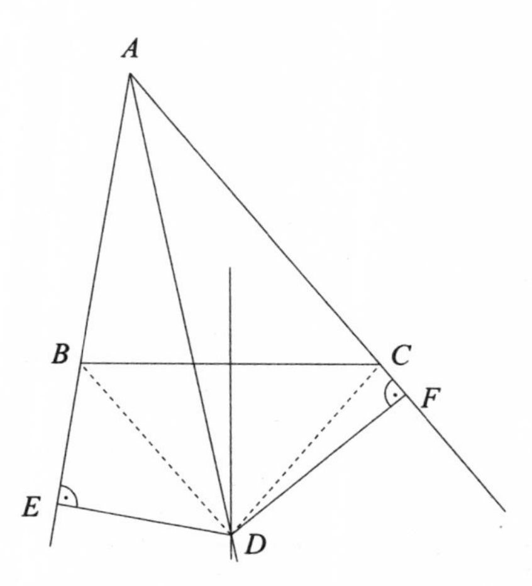

FIGURE 4.18. All triangles are equilateral! Better think twice!

Finally:

$$\left.\begin{array}{lcl} \triangle ADE \cong \triangle ADF & \Rightarrow & [AE] \cong [AF] \\ \triangle BDE \cong \triangle CDF & \Rightarrow & [BE] \cong [CF] \end{array}\right\} \Rightarrow [AB] \cong [AE] - [BE] \cong [AC]$$

If bisectors of angles $\angle A$ and side $[BC]$ intersect in the interior of $\triangle ABC$ and points E and F lie on segments $[AB]$ and $[AC]$, rather than on their extensions, the congruence of segments $[AB]$ and $[AC]$ is proved in a similar way by using difference of segments instead of their sum:

$$[AB] \cong [AE] + [BE] \cong [AF] + [CF] \cong [AC]$$

In any case we find $[AB] \cong [AC]$. Similarly we can show that also $[AB] \cong [BC]$; therefore $\triangle ABC$ is equilateral! □

We end our presentation here, although we have not mentioned many beautiful formulas and theorems about triangles, as well as circle and other plane figures. For example for convex quadrangles inscribed in a circle (cyclic quadrangles), there is a theorem due to Brahmagupta (seventh cent. A.D.):

$$P = \sqrt{(s-a)(s-b)(s-c)(s-d)}$$

where $s = (a+b+c+d)/2$.

Obviously for $d = 0$ we obtain Hero's formula.

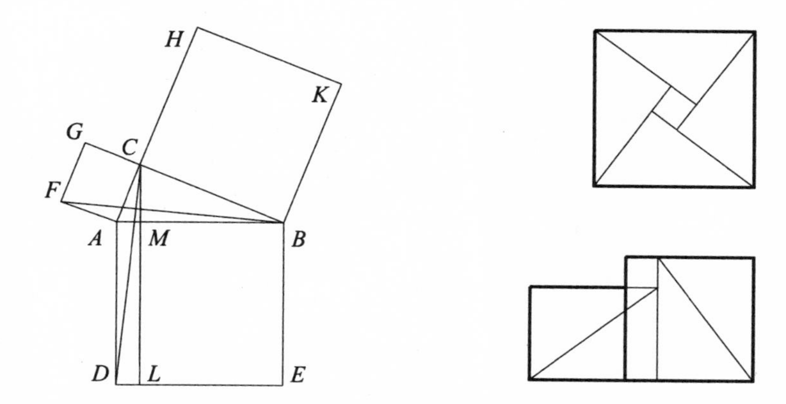

FIGURE 4.19. Proofs of the Pythagorean theorem by Euclid and Bhaskara.

4.4. Problems

In this section we present a few interesting and historically important examples. In addition, we give different proofs for some theorems proved earlier.

EXAMPLE 4.32 (EUCLID'S PROOF OF THE PYTHAGOREAN THEOREM). This is how the Pythagorean theorem was proved in the first book of Euclid's *Elements*:

In Fig. 4.19 we see that $P_{CAD} = P_{MAD} = \frac{1}{2}P_{MADL}$ because $\triangle CAD$ and $\triangle MAD$ are triangles with equal bases and altitudes. Similarly $P_{FAB} = P_{FAC} = \frac{1}{2}P_{FACG}$. But since $\triangle CAD \cong \triangle FAB$, we have $P_{MADL} = P_{FACG}$. Similarly $P_{MLEB} = P_{KHCB}$, hence:

$$P_{ADEB} = P_{MLEB} + P_{MADL} = P_{KHCB} + P_{FACG}$$

or expressed in the usual way:

$$c^2 = a^2 + b^2$$

EXAMPLE 4.33 (BHASKARA'S PROOF). Bhaskara, the twelfth-century Indian mathematician and astronomer, proved the Pythagorean theorem in one of his books by drawing a diagram (as on the right-hand side of Fig. 4.19) and by writing only *Behold!*

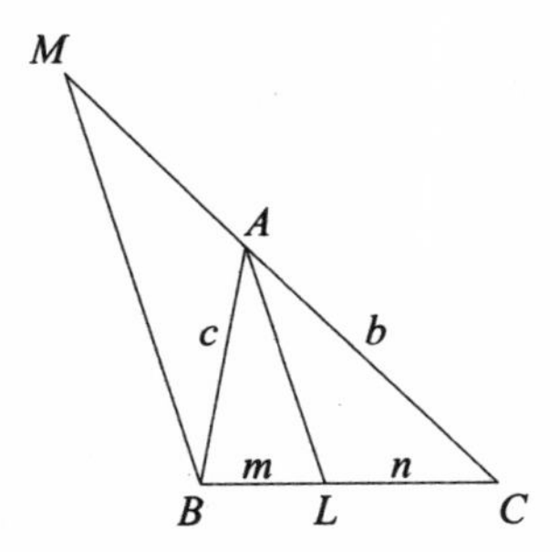

FIGURE 4.20. Angle bisectors.

EXAMPLE 4.34 (ANGLE BISECTOR REVISITED). Here is another proof of the theorem about the angle bisector (see Fig. 4.20).

Construct M on line AC such that $MB \parallel AL$. Then:

$$\left.\begin{array}{l}\angle MAB = 180^\circ - \alpha \\ \angle MBA = \alpha/2\end{array}\right\} \Rightarrow \angle BMA = \alpha/2 \Rightarrow \triangle AMB \text{ is isosceles}$$

Then $[AM] = c$. From $\triangle MBC \sim \triangle ALC$, we find

$$\frac{b}{n} = \frac{b+c}{m+n} \Rightarrow \frac{m+n}{n} = \frac{b+c}{b} \Rightarrow \frac{m}{n} = \frac{c}{b}$$

EXAMPLE 4.35 (CEVA'S THEOREM REVISITED). Draw a line $a \parallel BC$ through A that intersect lines at Q and R, respectively, as in Fig. 4.21. Then

$$\left.\begin{array}{r}\triangle BCY \sim \triangle QAY \Rightarrow \frac{[CY]}{[AY]} = \frac{[BC]}{[AQ]} \\ \triangle BCZ \sim \triangle ARZ \Rightarrow \frac{[AZ]}{[BZ]} = \frac{[AR]}{[BC]} \\ \left.\begin{array}{r}\triangle BXP \sim \triangle QAP \Rightarrow \frac{[BX]}{[AQ]} = \frac{[PX]}{[AP]} \\ \triangle CXP \sim \triangle RAP \Rightarrow \frac{[CX]}{[AR]} = \frac{[PX]}{[AP]}\end{array}\right\} \Rightarrow \frac{[BX]}{[CX]} = \frac{[AQ]}{[AR]}\end{array}\right\} \Rightarrow \frac{[BX]}{[CX]}\frac{[CY]}{[AY]}\frac{[AZ]}{[BZ]} = 1$$

EXAMPLE 4.36 (CONVERSE OF CEVA'S THEOREM). If we assume that:

$$\frac{[BX]}{[CX]}\frac{[CY]}{[AY]}\frac{[AZ]}{[BZ]} = 1$$

and that the intersection of line CP and side $[AB]$ is Z', where $\{P\} = [AX] \cap [BY]$, then according to Ceva's theorem:

$$\frac{[BX]}{[CX]}\frac{[CY]}{[AY]}\frac{[AZ']}{[BZ']} = 1$$

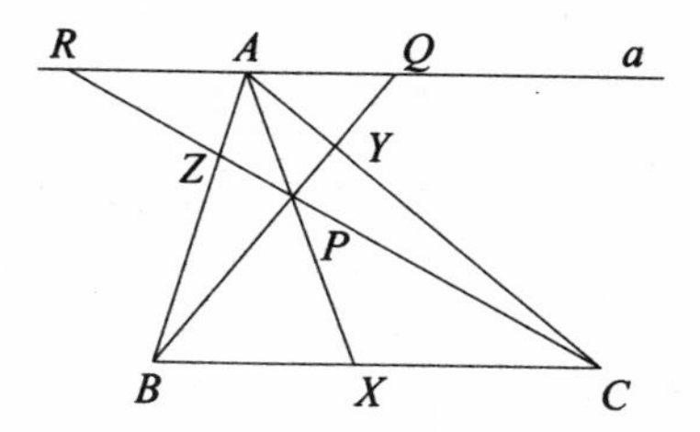

FIGURE 4.21. Ceva's theorem and its converse.

Therefore:

$$\frac{[AZ]}{[BZ]} = \frac{[AZ']}{[BZ']}$$

This implies $Z' \equiv Z$.

EXAMPLE 4.37 (GERGONNE'S POINT). Cevians defined by points where the incircle touches the triangle (see Fig. 4.22) satisfy the conditions of the converse of Ceva's theorem:

$$\frac{[BD]}{[CD]}\frac{[CE]}{[AE]}\frac{[AF]}{[BF]} = \frac{(s-b)(s-c)(s-a)}{(s-c)(s-a)(s-b)} = 1$$

Therefore they are concurrent. Their intersection is called *Gergonne's point*.

EXAMPLE 4.38 (CONSTRUCTION OF FERMAT'S POINT). Let us prove that Fermat's point can be constructed by first constructing the exterior equilateral triangles $\triangle XBC$, $\triangle AYC$, and $\triangle ABZ$ over the sides of $\triangle ABC$ and then the intersection of AX, BY, and CZ (see Fig. 4.22). In fact we prove here that these three lines

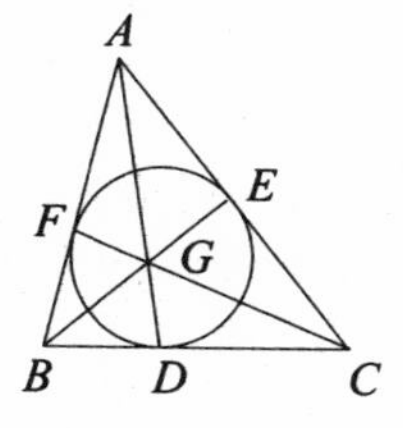

FIGURE 4.22. Gergonne's point.

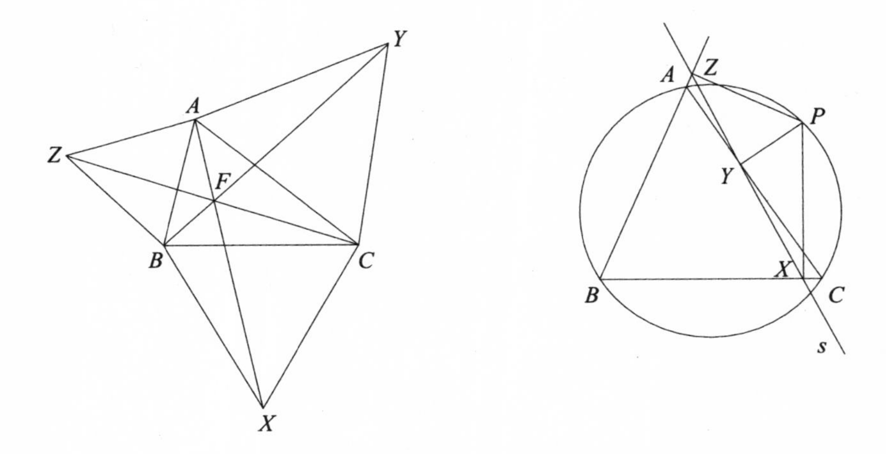

FIGURE 4.23. Fermat's point and Simson's line.

are concurrent (without Ceva's theorem). In the course of the proof, we see that their intersection is indeed Fermat's point.

Let F be defined by $\{F\} = BY \cap CZ$. Since the rotation of $\triangle AZC$ about A for 60° produces $\triangle ABY$, we find

$$\left.\begin{array}{l} \angle ZFB = 60^\circ \Rightarrow F \in \mathcal{K}_{AZB} \Rightarrow \angle AFZ = 60^\circ \\ \angle YFC = 60^\circ \Rightarrow F \in \mathcal{K}_{ACY} \Rightarrow \angle AFY = 60^\circ \end{array}\right\} \Rightarrow \angle BFC = 120^\circ$$

Therefore $F \in \mathcal{K}_{BXC}$, i.e., $\angle BFX = 60^\circ$, hence $F \in AX$. (See also Example 4.45.)

NOTE: The centers of the three external equilateral triangles form an equilateral triangle, the so-called *Napoleon's external triangle*. Similarly the centers of the internal equilateral triangles form *Napoleon's internal triangle*, which is also equilateral. The difference in the areas of Napoleon's external and internal triangles equals the area of the initial triangle $\triangle ABC$. Although Napoleon loved geometry, he probably did not have anything to do with these triangles.

EXAMPLE 4.39 (SIMSON'S LINE). The feet of perpendiculars from a point P on the circumcircle of $\triangle ABC$ on its sides are collinear, forming Simson's line (see Fig. 4.23).

SOLUTION: We prove the collinearity of feet of perpendiculars from P at AB, AC, and BC by showing that $\angle AYZ \cong \angle XYC$.

Since angles at X, Y, and Z are right, P lies on circumcircles of $\triangle XBZ$, $\triangle XYC$, and $\triangle AYZ$. Since also $\angle APC = 180° - \angle ABC$, we find

$$\angle APC \cong 180° - \angle ABC \cong 180° - \angle ZBX \cong \angle ZPX$$

Subtracting $\angle APX$ from the last equality yields $\angle XPC \cong \angle ZPA$.
Since the quadrangles $XYPC$ and $ZPYA$ are cyclic,* we have

$$\angle XYC = \angle XPC = \angle ZPA = \angle ZYA$$

Therefore X, Y, and Z are collinear. The line they lie on is called *Simson's line*.

EXAMPLE 4.40 (ARITHMETIC, GEOMETRIC, AND HARMONIC MEANS). In this example we present Pappus's proof of the inequality of the arithmetic, geometric, and harmonic means of two line segments or two numbers.

Let us see first where these three means can be found in a right triangle. Join the line segments x and y and construct a circle over their sum as a diameter. Then the radius of the circle is their arithmetic mean. In Fig. 4.24

$$[CO] = \frac{x+y}{2}$$

Since $\triangle ADC$ and $\triangle CDB$ are similar, we see that $[CD]$ is the geometric mean of x and y:

$$\frac{[AD]}{[CD]} = \frac{[CD]}{[BD]} \Rightarrow [CD] = \sqrt{xy}$$

If E is constructed as a foot of the perpendicular from D on $[CO]$, from the similarity of $\triangle CDO$ and $\triangle CED$, we find that $[CE]$ is the harmonic mean of x and y:

$$\frac{[CE]}{[CD]} = \frac{[CD]}{[CO]} \Rightarrow [CE] = \frac{2}{\frac{1}{x} + \frac{1}{y}}$$

In the following we prove the inequality of these three means. Since $[CO]$ is a hypotenuse of $\triangle CDO$, in which $[CD]$ is also a side, we find $[CO] \geq [CD]$,

*A quadrangle is cyclic if it is convex and a circle can be circumscribed about it.

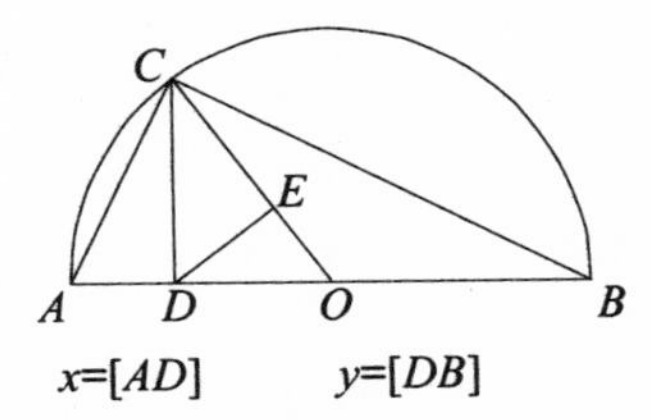

FIGURE 4.24. Pappus' proof of the inequality of the means.

i.e.:

$$\frac{x+y}{2} \geq \sqrt{xy}$$

Equality is achieved if and only if $D \equiv O$, i.e., if and only if $x = y$.

Similarly we find $[CD] \geq [CE]$, i.e.:

$$\sqrt{xy} \geq \frac{2}{\frac{1}{x}+\frac{1}{y}}$$

Again, equality is achieved if and only if $D \equiv O$, i.e., if and only if $x = y$.

Thus we proved that:

$$\frac{x+y}{2} \geq \sqrt{xy} \geq \frac{2}{\frac{1}{x}+\frac{1}{y}}$$

with the equalities if and only if $x = y$.

NOTE: For the proof of the more general inequality:

$$A_n \geq G_n \geq H_n$$

where A_n, G_n, and H_n are the arithmetic, geometric, and harmonic mean, respectively, of n positive numbers, see Appendix A, Example A.8.

EXAMPLE 4.41 (MEANS IN TRAPEZIUM). If the bases of a trapezium are $[AB] = x$ and $[CD] = y$ and the segments $[IJ]$, $[GH]$, and $[EF]$ are constructed so that I and J bisect the other two sides; G and H divide the other two sides so that the smaller trapeziums $ABHG$ and $GHCD$ are similar; E and F are their intersections with a line parallel to the bases, which passes through the

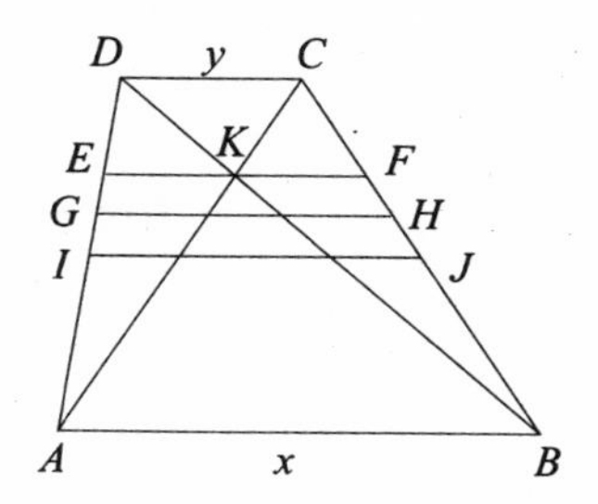

FIGURE 4.25. The means in a trapezium.

intersection of the diagonals (see Fig. 4.25); then:

$$[IJ] \text{ is the median line } \Rightarrow [IJ] = \frac{x+y}{2}$$

$$\frac{x}{[GH]} = \frac{[GH]}{y} \Rightarrow [GH]^2 = xy \Rightarrow [GH] = \sqrt{xy}$$

$$\left.\begin{array}{l} \frac{[ED]}{[EK]} = \frac{[AD]}{x} \\ \frac{[AE]}{[EK]} = \frac{[AD]}{y} \end{array}\right\} \Rightarrow \frac{[AD]}{[EK]} = \frac{[AD]}{x} + \frac{[AD]}{y} \Rightarrow [EF] = \frac{2}{\frac{1}{x} + \frac{1}{y}}$$

EXAMPLE 4.42 (AGAIN HERO'S FORMULA). Hero's proof of the area of a triangle in terms of its sides is much more geometric (see Fig. 4.26), although not entirely geometric, than the proof given in Example 4.2:

In Fig. 4.25, $[CH] = [AE]$, $\angle BIL = 90°$, and $\angle BCL = 90°$. Since $[CH] = [AE] = [AF]$, $[BD] = [BF]$, and $[CD] = [CE]$, we find $[BH] = s$ where

$$s = \frac{a+b+c}{2}$$

Recall (as in Example 4.14):

$$P_{ABC} = sr$$

Therefore:

$$(P_{ABC})^2 = [BH]^2[DI]^2$$

The following shows the similarity of $\triangle AIE$ and $\triangle BLC$, which finally leads us to the end of the proof:

$$\angle BIL \cong \angle BCL = 90° \Rightarrow \text{BICL is cyclic} \Rightarrow \angle BIC + \angle BLC = 180°$$

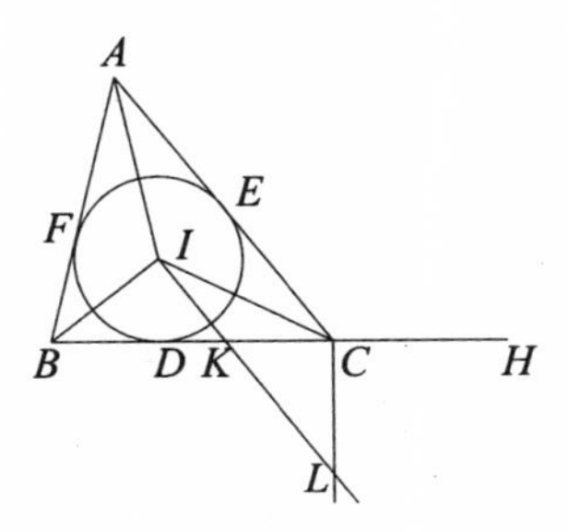

FIGURE 4.26. Hero's original proof of his formula.

Then:

$$(\triangle AFI \cong \triangle AEI,\ \triangle BDI \cong \triangle BFI,\ \triangle CEI \cong \triangle CDI) \Rightarrow \angle BIC + \angle AIE = 180^\circ$$

Therefore:

$$\angle AIE \cong \angle BLC$$

Since also $\angle AEI = \angle BCL = 90^\circ$, we find $\triangle AIE \sim \triangle BLC$. Finally:

$$\begin{aligned}\frac{[BC]}{[CL]} = \frac{[AE]}{[EI]} = \frac{[CH]}{[DI]} &\Rightarrow \frac{[BC]}{[CH]} = \frac{[CL]}{[DI]} = \frac{[CK]}{[DK]} \\ &\Rightarrow \frac{[BH]}{[CH]} = 1 + \frac{[BC]}{[CH]} = 1 + \frac{[CK]}{[DK]} = \frac{[CD]}{[DK]} \\ &\Rightarrow \frac{[BH]^2}{[BH][CH]} = \frac{[BD][CD]}{[BD][DK]} = \frac{[BD][CD]}{[DI]^2}\end{aligned}$$

because in the right triangle $\triangle BKI$, $[BD][DK] = [DI]^2$; hence:

$$(P_{ABC})^2 = [BH]^2[DI]^2 = [BH][CH][BD][CD] = s(s-a)(s-b)(s-c)$$

EXAMPLE 4.43 (POLYA'S PROBLEM). Let us prove that among all triangles with a given perimeter $O = 2s$ the largest area is covered by the equilateral triangle (whose sides are obviously $a = 2s/3$).

For an arbitrary triangle with a given perimeter O, i.e., semiperimeter s:

$$P_\triangle = \sqrt{s(s-a)(s-b)(s-c)} = \sqrt{s\left(\sqrt[3]{(s-a)(s-b)(s-c)}\right)^3}$$

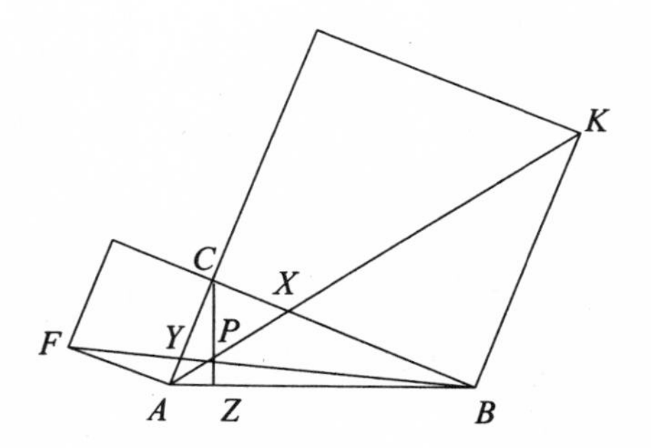

FIGURE 4.27. Application of the converse of Ceva's theorem.

Applying the inequality of the arithmetic and geometric means of three numbers, we obtain

$$P_{\triangle} \leq \sqrt{s\left(\frac{s-a+s-b+s-c}{3}\right)^3} = \sqrt{s\left(\frac{s}{3}\right)^3} = s^2\frac{\sqrt{3}}{9}$$

This equality holds if and only if $a = b = c$.

Let us briefly verify that expressions for the areas match:

$$s^2\frac{\sqrt{3}}{9} = \left(\frac{3a}{2}\right)^2\frac{\sqrt{3}}{9} = a^2\frac{\sqrt{3}}{4}$$

EXAMPLE 4.44 (FIGURE SIMILAR TO EUCLID'S). Let us prove that in a figure similar to the one Euclid used in his proof of the Pythagorean theorem, the lines AK and BF are concurrent with the altitude from C (see Fig. 4.27).

Indeed, using the converse Ceva's theorem:

$$\left.\begin{array}{lll}
\frac{[AZ]}{b} = \frac{b}{c} & \Rightarrow & [AZ] = \frac{b^2}{c} \\
\frac{[BZ]}{a} = \frac{a}{c} & \Rightarrow & [BZ] = \frac{a^2}{c} \\
\frac{[CX]}{a} = \frac{b}{a+b} & \Rightarrow & [CX] = \frac{ab}{a+b} \\
[BX] = a - [CX] & \Rightarrow & [BX] = \frac{a^2}{a+b} \\
\frac{[CY]}{b} = \frac{a}{a+b} & \Rightarrow & [CY] = \frac{ab}{a+b} \\
[AY] = b - [CY] & \Rightarrow & [AY] = \frac{b^2}{a+b}
\end{array}\right\} \Rightarrow \frac{[AZ]}{[BZ]}\frac{[BX]}{[CX]}\frac{[CY]}{[AY]} = 1$$

EXAMPLE 4.45 (GENERAL CASE OF FERMAT'S POINT). In this example we present the complete solution to Fermat's problem. The earlier solution is not satisfactory because if some of the angles of $\triangle ABC$ are $\geq 120°$, for example $\angle C$,

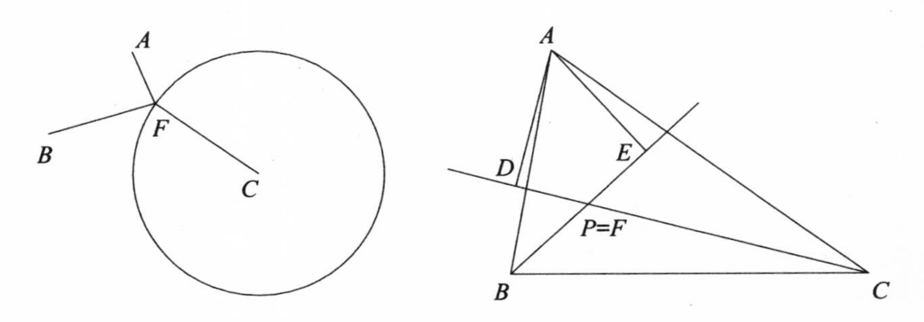

FIGURE 4.28. The general case of Fermat's point — the point that minimizes $[FA]+[FB]+[FC]$.

then Fermat's point coincides with that vertex. If all angles are $< 120°$, then Fermat's point can be found as before as the point that *sees* all sides at 120°.

Our goal is to find the point F for a given arbitrary $\triangle ABC$ such that the sum $[FA]+[FB]+[FC]$ achieves its minimum (see Fig. 4.28).

There are two possibilities:

- $F \in \{A,B,C\}$; i.e., F coincides with one of the vertices of $\triangle ABC$. In that case F coincides with the vertex of the greatest angle in the triangle, for example C. This is so because if $\angle C$ is the greatest angle, then $c = [AB]$ is the greatest side; hence $a+b < a+c$ and $a+b < b+c$.

- $F \notin \{A,B,C\}$, i.e., F does not coincide with any vertex of $\triangle ABC$. From the minimality requirement of the sum $[FA]+[FB]+[FC]$, we see that for any $[FC]$ such that vertices A and B are outside the circle $\mathcal{K}_C$, we must have $\angle AFC = \angle BFC$ (see Fig. 4.28). Similarly we find that $\angle AFB = \angle AFC$ must hold; therefore all three angles at F must equal 120°.

  To make this consideration complete, we must show that it is not possible that A or B is inside the circle $\mathcal{K}_C$. If at least one of them, e.g., A, is inside $\mathcal{K}_C$, then $[FC] \geq b$. Then from the triangle inequality, $[FA]+[FB] \geq c$. That implies $[FA]+[FB]+[FC] \geq b+c$, so that $F \equiv A$. This contradicts the assumption $F \notin \{A,B,C\}$. Similarly we find that A and C must lie outside $\mathcal{K}_B$, and that B and C must lie outside $\mathcal{K}_A$.

Let us now see which case is found for the given triangle $\triangle ABC$. In Example 4.38 we saw that if the point that sees all three sides at 120° exists it can be constructed as the intersection of circles circumscribed about external equilateral triangles. But if for example $\angle C \geq 120°$, then such a point does not exist; hence Fermat's point coincides with C, i.e., $F \equiv C$. If all angles of $\triangle ABC$ are $< 120°$, then the intersection P of these circles exists, and the only thing remaining to do is to show that if P exists, then $F \equiv P$.

We show that $[PA]+[PB]+[PC]$ is less than the sum of any two sides, for example $b+c$. Consider Fig. 4.28, where D is the perpendicular projection of A on CP and E is the perpendicular projection of A on BP. Hence $\angle APD = 60°$, implying that $[PD] = \frac{1}{2}[PA]$; therefore

$$b = [AC] > [CD] = [PC] + [PD] = [PC] + \frac{1}{2}[PA]$$

Similarly

$$c = [AB] > [BE] = [PB] + \frac{1}{2}[PA]$$

which implies $b+c > [PA]+[PB]+[PC]$.

We find similar relations for $a+b$ and $a+c$, so $F \notin \{A,B,C\}$, i.e., $F \equiv P$.

EXAMPLE 4.46 (POWER WITH RESPECT TO A CIRCLE). From an arbitrary point P in the plane of the given circle $\mathcal{K}$ draw a line intersecting $\mathcal{K}$, then denote the intersections by A and B. The product $[PA][PB]$ depends on only the distance of P from the center of the circle and the radius of the circle (see Fig. 4.29).

If we draw some other line through P, then denote its intersections with $\mathcal{K}$ by C and D, from $\triangle ACP \sim \triangle BDP$ (if P is inside $\mathcal{K}$), i.e., $\triangle BPC \sim \triangle APD$ (if P is outside $\mathcal{K}$), we find:

$$\frac{[PA]}{[PD]} = \frac{[PC]}{[PB]}$$

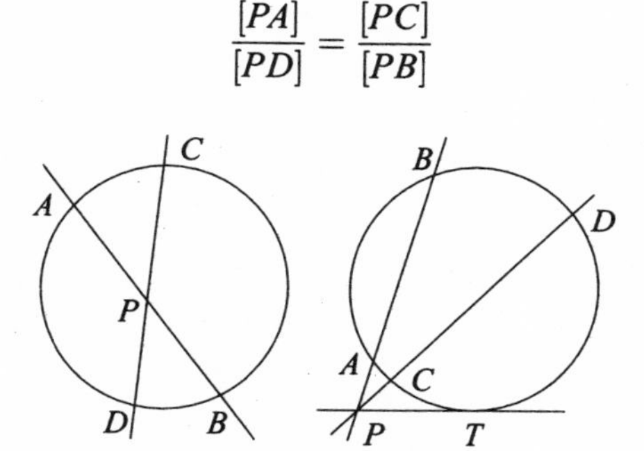

FIGURE 4.29. Power of a point with respect to a circle.

That is:

$$[PA][PB] = [PC][PD]$$

Thus we see that the product of the segments joining P and the intersections of the line through P with the circle $\mathcal{K}$ does not depend on the choice of the line. This product is called the *power of the point with respect to the circle*, and it is denoted by $\mathcal{P}_{\mathcal{K}}(P)$.

If P is outside $\mathcal{K}$, then in the special case when the line through P is tangent to $\mathcal{K}$, and therefore $A \equiv B \equiv T$, we find that $\mathcal{P}_{\mathcal{K}}(P) = [PT]^2$. Then obviously $\mathcal{P}_{\mathcal{K}}(P) = d^2 - r^2$, where $d = [PO]$ and r is the radius of the circle. The last formula also holds when P is inside the circle but with reversed signs: $\mathcal{P}_{\mathcal{K}}(P) = r^2 - d^2$. Often $\mathcal{P}_{\mathcal{K}}(P)$ is defined by $\mathcal{P}_{\mathcal{K}}(P) = d^2 - r^2$, so that the power of points inside the circle is negative.

EXAMPLE 4.47 (EULER'S FORMULA FOR $[OI]$ REVISITED). Let L be the intersection of the bisector of $\angle A$ and the circumcircle of $\triangle ABC$. Then L bisects the arch BC; hence the diameter LM is perpendicular to $[BC]$.

Observe that

$$\angle BML = \angle BAL = \frac{1}{2}\angle A \qquad \angle LBC = \angle LAC = \frac{1}{2}\angle A$$

Hence the external angle of $\triangle ABI$ at its vertex I equals

$$\angle BIL = \frac{1}{2}\angle A + \frac{1}{2}\angle B = \angle LBI$$

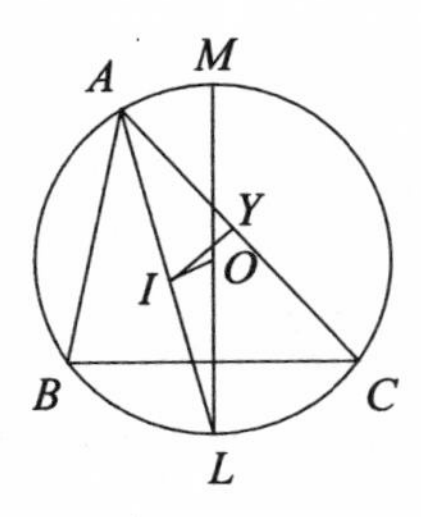

FIGURE 4.30. Euler's formula for $[OI]$.

Therefore $\triangle LBI$ is isosceles ($[LB] = [LI]$). Thus the power of I with respect to the circumcircle of $\triangle ABC$ is

$$\begin{aligned}\mathcal{P}_{\mathcal{K}}(I) &= R^2 - [OI]^2 = [LI][IA] \\ &= [LB][IA] = [LM]\frac{[LB]/[LM]}{[IY]/[IA]}[IY] \\ &= [LM][IY] = 2Rr\end{aligned}$$

Therefore we find the already familiar Euler's formula $[OI]^2 = R(R-2r)$.

EXAMPLE 4.48 (PTOLEMY'S THEOREM). If ABCD is a cyclic quadrangle, then

$$[AB][CD] + [AD][BC] = [AC][BD]$$

SOLUTION: Let $K \in [BD]$ be chosen so that $\angle BCK = \angle ACD$ (see Fig. 4.31). Then:

$$\left.\begin{aligned}&\left.\begin{aligned}\angle BCK &= \angle ACD \\ \angle CBD &= \angle CAD\end{aligned}\right\} \Rightarrow \triangle BCK \sim \triangle ACD \\ &\qquad\qquad\qquad \Rightarrow [AD][BC] = [AC][BK] \\ &\left.\begin{aligned}\angle KCD &= \angle ACB \\ \angle CDK &= \angle CAB\end{aligned}\right\} \Rightarrow \triangle CDK \sim \triangle ABC \\ &\qquad\qquad\qquad \Rightarrow [AB][CD] = [AC][KD]\end{aligned}\right\} \Rightarrow [AB][CD] + [AD][BC] = [AC][BD]$$

NOTE: This theorem was discovered by the Greek astronomer and mathematician Ptolemy of Alexandria (second cent. A.D.) and presented in his book *Almagest*, the astronomical encyclopedia used by the scientists until the seventeenth century. Ptolemy used this theorem to calculate the oldest known tables of trigonometric functions (also in *Almagest*).

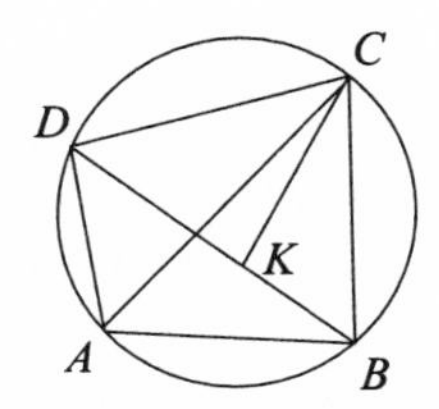

FIGURE 4.31. Ptolemy's theorem.

EXAMPLE 4.49 (EULER'S FORMULA). This formula due to Euler is certainly the most unusual formula in the entire field of mathematics:

$$e^{ix} = \cos x + i \sin x$$

Its rigorous proof is not possible without using complex analysis, the mathematical discipline developed in the nineteenth century. Here we present the plausibility argument, a nonrigorous derivation typical for Euler's age — the age of inspiration and discovery in analysis.

Write the Maclaurin series for e^{ix}, $\cos x$, and $\sin x$, then compare them:

$$\left.\begin{aligned} e^{ix} &= 1 + ix + \tfrac{(ix)^2}{2!} + \tfrac{(ix)^3}{3!} \\ &\quad + \tfrac{(ix)^4}{4!} + \tfrac{(ix)^5}{5!} + \dots \\ &= \left(1 - \tfrac{x^2}{2!} + \tfrac{x^4}{4!} - \dots\right) \\ &\quad + i\left(x - \tfrac{x^3}{3!} + \tfrac{x^5}{5!} - \dots\right) \\ \cos x &= 1 - \tfrac{x^2}{2!} + \tfrac{x^4}{4!} - \dots \\ \sin x &= x - \tfrac{x^3}{3!} + \tfrac{x^5}{5!} - \dots \end{aligned}\right\} \Rightarrow e^{ix} = \cos x + i \sin x$$

COROLLARY: For $x = \pi$ we obtain the relation between the five most important numbers in mathematics:

$$e^{i\pi} + 1 = 0$$

NOTE: The deficiency of this derivation is that, in Euler's time, the Maclaurin series was proved to work for real functions. The conditions for its extension to complex functions were found only later in the nineteenth century.

EXAMPLE 4.50 (SINE OF A SUM — GEOMETRICALLY). The easiest way of deriving trigonometric identities is by using Euler's formula:

$$e^{i\varphi} = \cos\varphi + i \sin\varphi$$

However let us recall their purely geometric derivation. For example let us derive the formula for the sine of a sum of two angles:

$$\begin{aligned} \sin(\alpha + \beta) &= \sin(180^\circ - \gamma) = \sin\gamma = \frac{c}{2R} = \frac{a\cos\beta + b\cos\alpha}{2R} \\ &= \sin\alpha\cos\beta + \sin\beta\cos\alpha \end{aligned}$$

EXAMPLE 4.51 (SUM OF SINES — ANALYTICALLY). Let us consider how Euler's formula is used to derive the formula for the sum of sines:

$$\begin{aligned}\sin\alpha+\sin\beta &= \frac{e^{i\alpha}-e^{-i\alpha}}{2i}+\frac{e^{i\beta}-e^{-i\beta}}{2i}\\ &= \frac{\left[e^{i(\alpha+\beta)/2}-e^{-i(\alpha+\beta)/2}\right]\left[e^{i(\alpha-\beta)/2}+e^{-i(\alpha-\beta)/2}\right]}{2i}\\ &= 2\sin\frac{\alpha+\beta}{2}\cos\frac{\alpha-\beta}{2}\end{aligned}$$

EXAMPLE 4.52 (HALF OF A GRAZING FIELD). A goat is tied to a fence around a circular field. If the radius of the field is r, what should be the length d of the rope, so that the goat can graze on exactly half of the field?

SOLUTION: This problem, so naively phrased, is well-known for the fact that it reduces to a transcendental equation that can be solved only by some iterative procedure.

One possible approach produces the following equation:

$$2\varphi\cos 2\varphi-\sin 2\varphi+\frac{\pi}{2}=0$$

where φ is the angle in radians, at the point where the rope is tied to the fence, between the diameter of a circle and the chord formed by the rope.

Solving this equation we obtain $\varphi = 0.9528$. Since $d = 2r\cos\varphi$, we find that:

$$d = 1.1587r$$

EXAMPLE 4.53 (GEOMETRY AND ALGEBRA). Algebraic equations are written using only algebraic operations: addition, subtraction, multiplication, division, powers, and roots. All other equations are called transcendental.

All algebraic and transcendental equations that have solution(s) can be solved to an arbitrary level of accuracy by using some iterative procedure. For scientists and engineers, a solution to two or three correct decimal places is often more than satisfactory. On the other hand for mathematicians, a more interesting question is whether or not exact solution of some equation can be found by radicals, i.e., by algebraic operations.

A general solution of the quadratic equation was known to al'Khwarizmi in the ninth cent. A.D. Formulas for solutions of cubic and quartic equations were found by Tartaglia and Ferrari in the sixteenth century, and first published by Cardano. All these formulas require only algebraic operations, so it was expected that the quintic and other equations would have similar solutions by radicals.

But although great effort was directed toward finding the solution for the quintic equation, no results were found until Abel showed in 1824 that there are quintics without a solution in radicals. The general criterion for solvability in radicals was found by Galois* in 1829.

As in the theory of equations, geometry has two different goals: to solve a problem in any way or to solve it using a ruler and compass only. The second approach is not necessarily the fastest or the simplest, but it is especially attractive because construction tools are so limited. Such solutions often have many steps; hence they are not very precise and primarily of theoretical interest.

Several geometric problems have remained unsolved since antiquity. We mention the four most famous: trisecting an angle,† doubling a cube,‡ squaring a circle§ (these are the three so-called *classical problems*), and constructing a regular heptagon.

Even ancient geometers knew how to solve some of these problems, but besides the straightedge and compass, they needed some other instruments. Hence the interest in such curves as the spiral of Archimedes, cissoid of Diocles, quadratrix of Dinostratus, and conchoid of Nicomedes.

When only nineteen Gauss used the analogy between constructions by straightedge and compass and the process of solving algebraic equations to show that the regular n-gon can be constructed if n is a product of different Fermat primes¶ or a unity with a nonnegative power of two. Therefore we can construct regular polygons for

$$n = 2,3,4,5,6,8,10,12,15,16,17,20,\ldots$$

*Evarist Galois had a very tragic destiny — he was killed in a duel at twenty-one. Since he had unsuccessfully tried to publish his results during the previous two years, and he was aware of the impending tragedy, the night before the duel, he wrote a letter to his friend, Auguste Chevalier, presenting the elements of his theory. His work was published 14 years after his death, in 1846.

†The problem is to trisect an arbitrary angle.

‡This problem is also known as the Delian problem, because, by the legend, the oracle at Delos advised the Athenians to double the cubical altar of Apollo in order to stop the plague of 430 B.C.

§The problem here is to construct the square of the area equal to the given circle.

¶Of all Fermat numbers $F_k = 2^{2^k} + 1$ $(k = 0, 1, 2, \ldots)$ it is quite probable that only $F_0 = 3$, $F_1 = 5$, $F_2 = 17$, $F_3 = 257$, and $F_4 = 65537$ are prime, but that is still an unsolved problem.

Although Gauss probably knew that this condition is not only sufficient but also necessary, this part of the solution was published later. Thus construction is impossible for

$$n = 7, 9, 11, 13, 14, 18, 19, 21, 22, 23, \ldots$$

The other problems reduce to questions of the constructibility of the solution of a cubic equation (angle trisection and cube duplication), and the number π (squaring the circle). In the seventeenth century Descartes observed that Euclidean constructions (those using only the straightedge and compass) allow us to construct only solutions of linear and quadratic equations; i.e., the construction of rational numbers and the quadratic irrationals only. Hence square roots can be constructed but third roots cannot; therefore angle trisection and the Delian problem cannot be solved by Euclidean constructions. The impossibility to construct the number π followed from Lindemann's proof of its transcendence* in 1882.

---

*A number is transcendental if it is not a root of an algebraic equation.

Appendixes

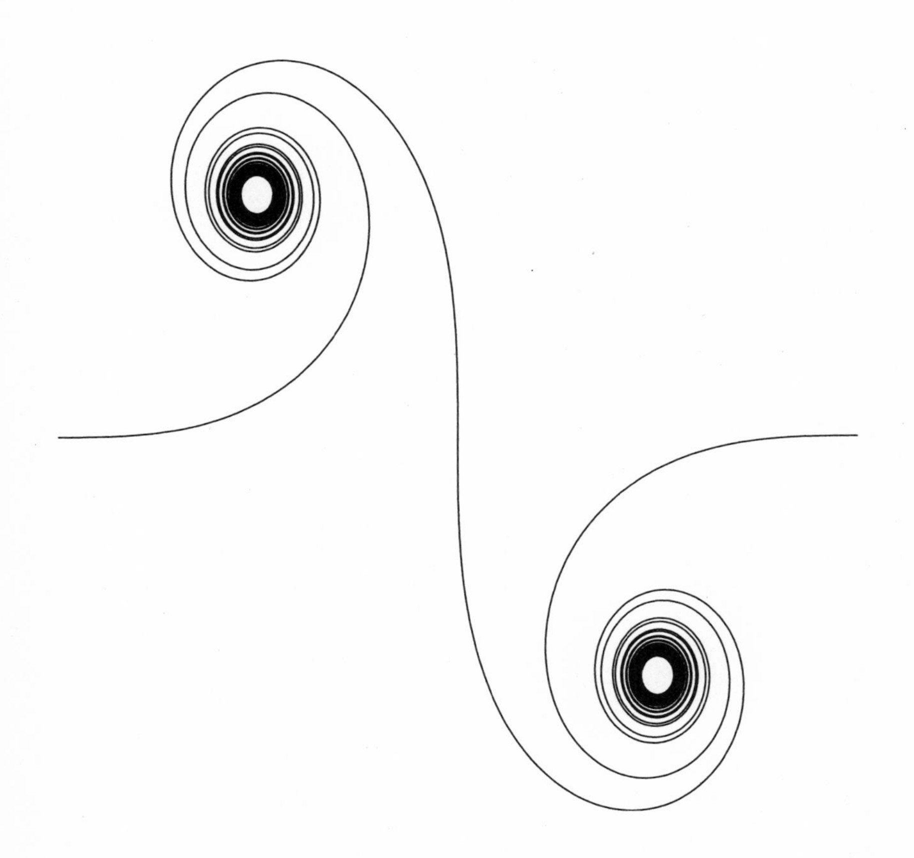

Appendix A

Mathematical Induction

Appendix A presents an important mathematical tool, *mathematical induction*, primarily through examples.

A.1. Overview

In many mathematical problems we want to prove that the statement A_n is true for any value of the natural number $n \geq n_{min}$. If the statement we are trying to prove is indeed true, very often we can use the method of mathematical induction to facilitate the proof. We need to show that:

1. The statement is true for $n = n_{min}$.
2. From the correctness of the statement for $n = k$ $(k \geq n_{min})$, it follows that the statement is true for $n = k + 1$.

Historians of mathematics have different opinions about who first formulated the principle of mathematical induction. However it is certain that Ancient Greek mathematicians used it. For example Theorem IX-20 in Euclid's *Elements* proves the infiniteness of the set of primes by showing that if there are n primes, there must exist the $(n+1)$th prime, too.

For the precise formulation of this theorem, we should probably credit Jakob Bernoulli and Blaise Pascal. In 1889 Giuseppe Peano introduced the principle of mathematical induction among the axioms of natural numbers.

A proof by mathematical induction can be imagined as a row of falling dominoes. To ensure that all dominoes fall, first we verify that the first domino can fall. After that if for every $k \geq 1$ we can prove that the fall of the kth domino produces the fall of the next, i.e., the $(k+1)$th domino, we can be certain that all dominoes will fall.

We continue with a few applications of the principle of mathematical induction.

A.2. Examples

EXAMPLE A.1 (SUM OF FIRST n NATURAL NUMBERS). Let us prove that for all $n \in N$:

$$1+2+3+\ldots+n = \frac{n(n+1)}{2}$$

Step 1: Check for $n = 1$

$$1 = \frac{1 \cdot 2}{2} \qquad \checkmark$$

Step 2: We assume that for $n = k$, $k \geq 1$ the formula is true, then try to prove that this implies its correctness for $n = k+1$:

$$\begin{aligned} 1+2+3+\ldots+k+(k+1) &= \frac{k(k+1)}{2} + (k+1) \\ &= \frac{(k+1)(k+2)}{2} \end{aligned}$$

This completes the proof.

EXAMPLE A.2 (SUM OF FIRST n SQUARES). Here we prove that for all $n \in N$:

$$1^2 + 2^2 + 3^2 + \ldots + n^2 = \frac{n(n+1)(2n+1)}{6}$$

Step 1: Check for $n = 1$

$$1^2 = \frac{1 \cdot 2 \cdot 3}{6} \qquad \checkmark$$

Step 2: If the statement is true for $n = k$, $k \geq 1$, then for $n = k+1$:

$$\begin{aligned} 1^2 + 2^2 + 3^2 + \ldots + k^2 + (k+1)^2 &= \frac{k(k+1)(2k+1)}{6} + (k+1)^2 \\ &= \frac{(k+1)(k+2)(2k+3)}{6} \end{aligned}$$

This completes the proof.

EXAMPLE A.3 (SUM OF FIRST n CUBES). Let us show that for all $n \in N$:

$$1^3 + 2^3 + 3^3 + \ldots + n^3 = \left[\frac{n(n+1)}{2}\right]^2$$

Step 1: Check for $n = 1$:

$$1^3 = \left(\frac{1 \cdot 2}{2}\right)^2 \qquad \checkmark$$

Step 2: Let the formula be true for $n = k$, $k \geq 1$. Then for $n = k+1$:

$$1^3 + 2^3 + 3^3 + \ldots + k^3 + (k+1)^3 = \left[\frac{k(k+1)}{2}\right]^2 + (k+1)^3$$
$$= \left[\frac{(k+1)(k+2)}{2}\right]^2$$

This completes the proof.

NOTE: By combining this result and the identity $1+2+3+\ldots+n = n(n+1)/2$, we obtain:

$$(1+2+3+\ldots+n)^2 = 1^3 + 2^3 + 3^3 + \ldots + n^3$$

See Example 2.52 for the combinatorial proof of this identity. □

Is there a simple method of finding sums similar to those in previous examples or such sums as:

$$\sum_{i=1}^{n}(2i-1) \quad \text{or} \quad \sum_{i=1}^{n}(2i-1)^2$$

especially if we do not know (or forget) that:

$$\sum_{i=1}^{n}(2i-1) = n^2 \quad \text{and} \quad \sum_{i=1}^{n}(2i-1)^2 = \frac{n(2n-1)(2n+1)}{3}$$

Indeed when we know where to begin, using mathematical induction to prove such identities is a routine matter.

The identity $1+2+\ldots+n = n(n+1)/2$ can be found using the same idea that Gauss had when he was only nine years old and impressed his teacher by quickly summing $1+2+\ldots+100 = 5050$.

Indeed:

$$1+2+\ldots+99+100 = (1+100)+(2+99)+\ldots+(50+51)$$
$$= \frac{100}{2} \cdot 101$$

The solution to the sum of the first n odd numbers, i.e., $1+3+5+\ldots+(2n-1) = n^2$, is clear if we draw a square 5×5 (Fig. A.1), then note that it is made

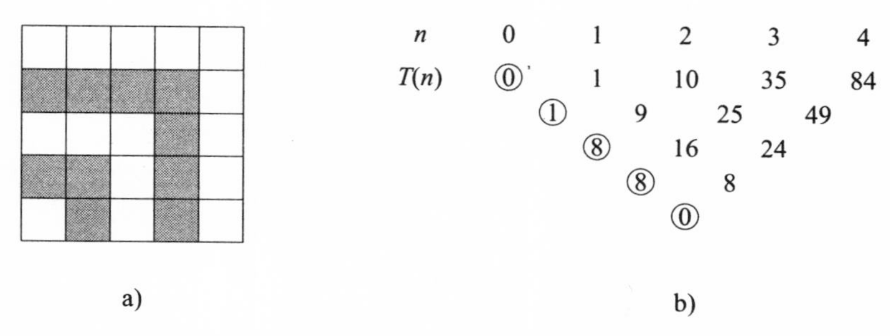

FIGURE A.1. a) An illustration of $1+3+5+7+9=5^2$. b) Gregory's triangle for $T(n)=1^2+3^2+\ldots+(2n-1)^2$.

up of 1, 3, 5, 7, and 9 unit squares. In general we notice that the difference of areas of squares $n \times n$ and $(n-1) \times (n-1)$ equals the nth odd number:

$$n^2-(n-1)^2=2n-1$$

For other sums the guessing process is not so simple. In such cases Gregory's triangle is often useful.

EXAMPLE A.4 (GREGORY'S TRIANGLE). Let us take a look at how Gregory's triangle can be used to determine the sum of squares of the first n odd numbers:

$$T(n)=\sum_{i=1}^{n}(2i-1)^2=1^2+3^2+\ldots+(2n-1)^2$$

First we calculate $T(n)$ for $n=0,1,2,3,4,\ldots$ Note: $T(0)=0$ because the empty sum by definition equals zero just as the empty product is always one. (Recall the way we initialize sums and products in computer programs before entering a loop.) Next we form a table, like the one in Fig. A.1. The first row consists of calculated values of $T(n)$, the second row are differences of elements in the first row, and so on until we reach a row of all zeros.

The circled numbers $0,1,8,8,0$, which we call the Gregory transformation of the sequence $T(n)$, we then multiply by the binomial coefficients:

$$\binom{n}{0} \quad \binom{n}{1} \quad \binom{n}{2} \quad \binom{n}{3} \quad \binom{n}{4} \quad \ldots$$

In this way we find $T(n)$ in a closed form:

$$\begin{aligned} T(n) &= 0 \cdot \binom{n}{0} + 1 \cdot \binom{n}{1} + 8 \cdot \binom{n}{2} + 8 \cdot \binom{n}{3} + 0 \cdot \binom{n}{4} \\ &= 0 \cdot 1 + 1 \cdot n + 8 \cdot \frac{n(n-1)}{2} + 8 \cdot \frac{n(n-1)(n-2)}{6} \\ &= \frac{n(2n-1)(2n+1)}{3} \end{aligned}$$

It is left to the reader to prove this identity by induction.

NOTE: No matter how we guess at an identity or a theorem, mere guessing is not enough — we must always have an adequate proof. Many cases illustrate this point, for example when Fermat found that the numbers:

$$F_n = 2^{2^n} + 1 \qquad n = 0, 1, \ldots$$

are prime for $n = 0, 1, 2, 3$, and 4, he conjectured that all numbers F_n are prime. But Euler later showed that $F_5 = 4294967297 = 641 \cdot 6700417$. It is still not known if there are other primes among the Fermat numbers except for those found by Fermat himself, i.e., F_0, F_1, F_2, F_3, and F_4.

EXAMPLE A.5 (EASY MATRIX). Consider the nth power of the matrix:

$$A = \begin{bmatrix} 1 & 1 \\ 0 & 1 \end{bmatrix}$$

Let us first calculate A^n for $(n = 2, 3, 4)$:

$$A^2 = \begin{bmatrix} 1 & 2 \\ 0 & 1 \end{bmatrix} \qquad A^3 = \begin{bmatrix} 1 & 3 \\ 0 & 1 \end{bmatrix} \qquad A^4 = \begin{bmatrix} 1 & 4 \\ 0 & 1 \end{bmatrix}$$

In this case regularity is easily observable, so it is probably true that:

$$A^n = \begin{bmatrix} 1 & n \\ 0 & 1 \end{bmatrix}$$

Hence without further delay, we proceed with the inductive proof.

Step 1:

$$\begin{bmatrix} 1 & 1 \\ 0 & 1 \end{bmatrix}^1 = \begin{bmatrix} 1 & 1 \\ 0 & 1 \end{bmatrix}. \qquad \checkmark$$

Step 2: Suppose:

$$A^k = \begin{bmatrix} 1 & k \\ 0 & 1 \end{bmatrix}$$

Then

$$A^{k+1} = A^k \cdot A = \begin{bmatrix} 1 & k \\ 0 & 1 \end{bmatrix} \cdot \begin{bmatrix} 1 & 1 \\ 0 & 1 \end{bmatrix} = \begin{bmatrix} 1 & k+1 \\ 0 & 1 \end{bmatrix}$$

This completes the proof.

EXAMPLE A.6 (VANDERMONDE DETERMINANT). The Vandermonde determinant of order n is defined by:

$$V_n(a_1,\ldots,a_n) = \begin{vmatrix} 1 & a_1 & \ldots & a_1^{n-1} \\ 1 & a_2 & \ldots & a_2^{n-1} \\ \vdots & & & \\ 1 & a_n & \ldots & a_n^{n-1} \end{vmatrix}$$

We use induction to prove that for $n \geq 2$:

$$V_n(a_1,\ldots,a_n) = \prod_{1 \leq i < j \leq n} (a_j - a_i)$$

For example for $n = 3$ we find

$$V_3(a,b,c) = \begin{vmatrix} 1 & a & a^2 \\ 1 & b & b^2 \\ 1 & c & c^2 \end{vmatrix} = (c-a)(c-b)(b-a)$$

Step 1: For $n = 2$ we have

$$\begin{vmatrix} 1 & a_1 \\ 1 & a_2 \end{vmatrix} = a_2 - a_1. \qquad \checkmark$$

Step 2: Let

$$V_k(a_1,\ldots,a_k) = \prod_{1 \leq i < j \leq k} (a_j - a_i)$$

If in the determinant:

$$V_{k+1}(a_1,\ldots,a_{k+1}) = \begin{vmatrix} 1 & a_1 & a_1^2 & \ldots & a_1^k \\ 1 & a_2 & a_2^2 & \ldots & a_2^k \\ \vdots & & & & \\ 1 & a_k & a_k^2 & \ldots & a_k^k \\ 1 & a_{k+1} & a_{k+1}^2 & \ldots & a_{k+1}^k \end{vmatrix}$$

from the jth column we subtract the $(j-1)$th column multiplied by a_1, for all $j = 2,3,\ldots,(k+1)$, and then extract $(a_i - a_1)$ from the ith row, for every $i = 2,3,\ldots,(k+1)$, we find

$$V_{k+1}(a_1,\ldots,a_{k+1}) = \begin{vmatrix} 1 & 0 & 0 & \ldots & 0 \\ 1 & 1 & a_2 & \ldots & a_2^{k-1} \\ \vdots & & & & \\ 1 & 1 & a_k & \ldots & a_k^{k-1} \\ 1 & 1 & a_{k+1} & \ldots & a_{k+1}^{k-1} \end{vmatrix} \cdot \prod_{i=2}^{k+1}(a_i - a_1)$$

Using Laplace's determinant expansion, we find

$$V_{k+1}(a_1,\ldots,a_{k+1}) = 1 \cdot V_k(a_2,\ldots,a_{k+1}) \cdot \prod_{j=2}^{k+1}(a_j - a_1)$$

Then using the inductive hypothesis:

$$V_{k+1}(a_1,\ldots,a_{k+1}) = \prod_{1 \le i < j \le k+1}(a_j - a_i)$$

This completes the proof.

EXAMPLE A.7 (PLANE AND n LINES). We are given $n \geq 0$ lines having an arbitrary arrangement in the plane. Let us find $L(n)$, the number of regions these n lines define in the plane.

We can use Gregory's triangle (see Fig. A.2) to guess the formula for $L(n)$: We find that probably:

$$L(n) = 1 \cdot \binom{n}{0} + 1 \cdot \binom{n}{1} + 1 \cdot \binom{n}{2} = \frac{n(n+1)}{2} + 1$$

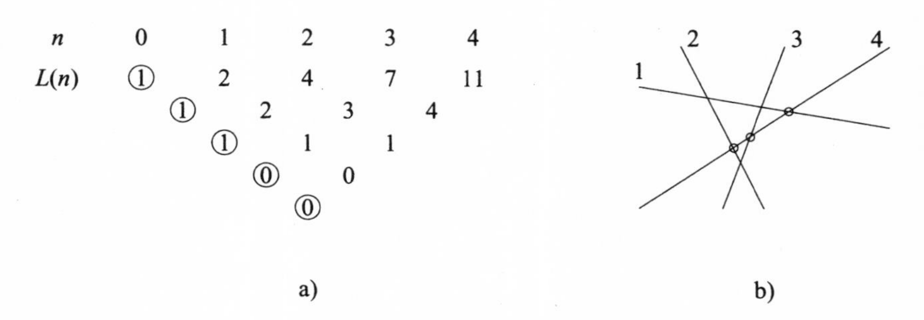

FIGURE A.2. a) Gregory's triangle for $L(n)$. b) The transition from $n=3$ to $n=4$.

Let us use induction:

Step 1: For $n = n_{min} = 0$:

$$1 = \frac{0 \cdot 1}{2} + 1. \qquad \checkmark$$

Step 2: Suppose:

$$L(k) = \frac{k(k+1)}{2} + 1$$

Since:

$$L(k+1) = L(k) + k + 1$$

When we draw the $(k+1)$th line, it intersects the others at k points and adds $(k+1)$ new regions to the division (see Fig. A.2); therefore:

$$\begin{aligned} L(k+1) &= \frac{k(k+1)}{2} + 1 + k + 1 \\ &= \frac{(k+1)(k+2)}{2} + 1 \end{aligned}$$

This completes the proof.

EXAMPLE A.8 (ARITHMETIC, GEOMETRIC, AND HARMONIC MEANS). Let us prove that for every $n > 1$ and arbitrary positive numbers x_i $(i = 1, 2, \ldots, n)$ the following inequalities hold

$$A_n \geq G_n \geq H_n$$

where:

$$A_n = \frac{x_1 + x_2 + \ldots + x_n}{n} \qquad G_n = \sqrt[n]{x_1 x_2 \ldots x_n} \qquad H_n = \frac{n}{\frac{1}{x_1} + \frac{1}{x_2} + \ldots + \frac{1}{x_n}}$$

are the arithmetic, geometric, and harmonic means of numbers x_i $(i = 1, 2, \ldots, n)$, respectively.

We first prove that the inequality $A_n \geq G_n$ is true for infinitely many values of n: Step 1: For $n = 2$ we have

$$(x_1 - x_2)^2 \geq 0 \Rightarrow \frac{x_1 + x_2}{2} \geq \sqrt{x_1 x_2}. \qquad \checkmark$$

Step 2: Assume $A_k \geq G_k$, then:

$$A_{2k} = \frac{x_1 + \ldots + x_{2k}}{2k} = \frac{(x_1 + \ldots + x_k)/k + (x_{k+1} + \ldots + x_{2k})/k}{2}$$

$$A_{2k} \geq \frac{\sqrt[k]{x_1 \ldots x_k} + \sqrt[k]{x_{k+1} \ldots x_{2k}}}{2} \geq \sqrt{\sqrt[k]{x_1 \ldots x_k}\,\sqrt[k]{x_{k+1} \ldots x_{2k}}}$$

$$A_{2k} \geq \sqrt[2k]{x_1 \ldots x_k x_{k+1} \ldots x_{2k}} = G_{2k}$$

Thus $A_n \geq G_n$ is true for for $n = 2, 4, 8, 16, \ldots$

Second we prove that for $n > 2$:

$$A_n \geq G_n \Rightarrow A_{n-1} \geq G_{n-1}$$

Indeed if:

$$\frac{x_1 + x_2 + \ldots + x_n}{n} \geq \sqrt[n]{x_1 x_2 \ldots x_n}$$

holds true for arbitrary positive numbers x_i $(i = 1, 2, \ldots, n)$, then it must be true for the special case when:

$$x_n = \frac{x_1 + x_2 + \ldots + x_{n-1}}{n-1}$$

That is:

$$\frac{x_1 + \ldots + x_{n-1} + \frac{x_1 + x_2 + \ldots + x_{n-1}}{n-1}}{n} \geq \sqrt[n]{x_1 \ldots x_{n-1} \frac{x_1 + x_2 + \ldots + x_{n-1}}{n-1}}$$

which yields

$$\frac{x_1 + x_2 + \ldots + x_{n-1}}{n-1} \geq \sqrt[n-1]{x_1 x_2 \ldots x_{n-1}}$$

Thus we showed that:

$$A_n \geq G_n \quad (n > 1)$$

The other part $(G_n \geq H_n)$ is a trivial consequence of the preceding inequality. This completes the proof.

NOTES: Both inequalities become equalities if and only if all numbers x_i $(i = 1, 2, \ldots, n)$ are equal. (See Problem A.17.) The technique applied in this proof is called *regressive induction*. The preceding proof was given by Cauchy in 1821. The inequality $A_2 \geq G_2$ was known to the Ancient Greeks. The first proof of the general inequality was given by Maclaurin around 1729. The geometric proof of the case $n = 2$ is found in Example 4.40. □

References [3] and [28] give examples of theorems that cannot be proved by induction, even though more general theorems are easily proved by that same technique. Here we present one of two examples from Ref. [3]. The other example is found in Problem A.18 at the end of this appendix.

EXAMPLE A.9 (MORE GENERAL IS SOMETIMES EASIER). We prove that for $n > 1$:

$$\frac{1}{2^2} + \frac{1}{3^2} + \ldots + \frac{1}{n^2} < 1 - \frac{1}{n} \tag{A.1}$$

Step 1: For $n = n_{min} = 2$ it is obvious that:

$$\frac{1}{2^2} < 1 - \frac{1}{2}. \qquad \checkmark$$

Step 2: Assume that:

$$\frac{1}{2^2} + \frac{1}{3^2} + \ldots + \frac{1}{k^2} < 1 - \frac{1}{k}$$

Then:

$$\frac{1}{2^2} + \frac{1}{3^2} + \ldots + \frac{1}{k^2} + \frac{1}{(k+1)^2} < 1 - \frac{1}{k} + \frac{1}{(k+1)^2}$$

If:

$$1 - \frac{1}{k} + \frac{1}{(k+1)^2} \stackrel{?}{\geq} 1 - \frac{1}{k+1}$$

it implies

$$k^2 + 2k \stackrel{?}{\geq} (k+1)^2$$

This is false; therefore:

$$1 - \frac{1}{k} + \frac{1}{(k+1)^2} < 1 - \frac{1}{k+1}$$

That is:

$$\frac{1}{2^2} + \frac{1}{3^2} + \ldots + \frac{1}{k^2} + \frac{1}{(k+1)^2} < 1 - \frac{1}{k+1}$$

This completes the proof.

NOTE: If we try to prove by induction that

$$\frac{1}{2^2} + \frac{1}{3^2} + \ldots + \frac{1}{n^2} < 1$$

the inequality which is an immediate consequence of Eq. (A.1) and $1 - \frac{1}{n} < 1$, we find unsurmountable difficulties in going from $n = k$ to $n = k+1$.

G. Polya explains this apparent paradox in Ref. [45] as follows:

> In general, in trying to devise a proof by mathematical induction, you may fail for two opposite reasons. You may fail because you try to prove too much: Your A_{n+1} is too heavy a burden. Yet you may also fail because you try to prove too little: Your A_n is too weak a support. In general, you have to balance the statement of your theorem so that the support is just enough for the burden.

A.3. Problems

PROBLEM A.1. In 1772 Euler found that the quadratic trinomial:

$$n^2 + n + 41$$

produces different primes for $n = 0, 1, 2, \ldots, 39$. Prove that its value for $n = 40$ is a composite number.

PROBLEM A.2. Determine the closed form expressions for:

a. $R_1(n) = 1 + 2 + \ldots + n$

b. $R_2(n) = 1 \cdot 2 + 2 \cdot 3 + \ldots + n(n+1)$

c. $R_3(n) = 1 \cdot 2 \cdot 3 + 2 \cdot 3 \cdot 4 + \ldots + n(n+1)(n+2)$

d. Generalize the cases a–c.

PROBLEM A.3. Find the closed form expression for:

$$S_4(n) = 1^4 + 2^4 + \ldots + n^4$$

PROBLEM A.4. Determine the expression for the sum of the first n members of an arithmetic series:

$$A(n) = a + (a+d) + (a+2d) + \ldots + [a + (n-1)d]$$

PROBLEM A.5. Find the expression for the sum of the first n numbers from a geometric series:

$$B(n) = b + bq + bq^2 + \ldots + bq^{n-1}$$

PROBLEM A.6. Show that:

$$\frac{1}{1 \cdot 2} + \frac{1}{2 \cdot 3} + \frac{1}{3 \cdot 4} + \ldots + \frac{1}{n(n+1)} = \frac{n}{n+1}$$

PROBLEM A.7. Prove that:

$$\left(1 - \frac{1}{4}\right)\left(1 - \frac{1}{9}\right) \ldots \left(1 - \frac{1}{n^2}\right) = \frac{n+1}{2n}$$

PROBLEM A.8. Prove Newton's binomial formula (for the combinatorial proof see Example 2.27)

$$(a+b)^n = \sum_{i=0}^{n} \binom{n}{i} a^{n-i} b^i \qquad n \in N$$

PROBLEM A.9. Define the Euclid numbers by:

$$e_n = e_1 e_2 \ldots e_{n-1} + 1 \qquad e_1 = 2$$

a. Are all Euclid numbers prime?

b. Determine

$$\sum_{i=1}^{n} \frac{1}{e_i}$$

c. If $p_1, p_2, \ldots, p_n$ are the first n primes, is $p_1 p_2 \ldots p_n + 1$ a prime for every $n \in N$?

d. Define numbers c_n by the following recursion:

$$c_{n+1} = 2^{c_n} - 1 \qquad c_0 = 2$$

Cantor's conjecture states that *All numbers c_n are prime.* Calculate several of the numbers c_n.

PROBLEM A.10. Prove that $3 \cdot 4^{n+1} + 10^{n-1} - 4 \quad (n \in N)$ is divisible by 9.

PROBLEM A.11. Show that for the Fibonacci numbers Binet's formula holds

$$f_n = \frac{\sqrt{5}}{5}\left[\left(\frac{1+\sqrt{5}}{2}\right)^n - \left(\frac{1-\sqrt{5}}{2}\right)^n\right]$$

PROBLEM A.12. Find A^n if:

$$A = \begin{bmatrix} 1 & 1 \\ 1 & 0 \end{bmatrix}$$

PROBLEM A.13. Prove the de Moivre's formula:

$$(\cos\varphi + i\sin\varphi)^n = \cos n\varphi + i\sin n\varphi \qquad n \in N$$

PROBLEM A.14. Define $T_n(x)$ for $x \in [-1,1]$ and $n \in N_0$ as follows:

$$T_n(x) = \cos(n \cdot \arccos x)$$

Prove that on the segment $[-1,1]$ the function $T_n(x)$ can be represented as an order-n polynomial.

NOTE: Polynomials $T_n(x)$ are called Chebyshev polynomials.

PROBLEM A.15. Prove that for any $n \in N$:

$$2\cos(\pi/2^{n+1}) = \sqrt{2+\sqrt{2+\ldots+\sqrt{2}}}$$

On the right-hand side, there are n nested square roots.

PROBLEM A.16. Find the Maclaurin expansion of $f(x) = \ln(1+x)$. Recall that the Maclaurin series has the following form:

$$\sum_{n=0}^{\infty} \frac{f^{(n)}(0)}{n!} x^n$$

PROBLEM A.17. Use mathematical induction to prove that for arbitrary positive numbers $a_1, a_2, \ldots, a_n$

$$a_1 a_2 \ldots a_n = 1 \;\Rightarrow\; a_1 + a_2 + \ldots + a_n \geq n$$

Use this inequality to prove the inequality of the arithmetic, geometric, and harmonic means.

PROBLEM A.18. Use the induction to prove that:

$$\frac{1}{2} \cdot \frac{3}{4} \cdot \ldots \cdot \frac{2n-1}{2n} < \frac{1}{\sqrt{3n}}$$

NOTE: In this case it is easier to prove a more restrictive inequality.

A.4. Hints and Notes

HINT A.1. Indeed, $40^2 + 40 + 41 = 41^2$. Euler's trinomial produces an incredibly long sequence of primes, which could make us think that all of its values are prime. The following anecdote from Ref. [45] depicts different inductive reasonings:

> "Look at this mathematician," said the logician. "He observes that the first ninety-nine numbers are less than hundred and infers hence, by what he calls induction, that all numbers are less than a hundred."
>
> "A physicist believes," said the mathematician, "that 60 is divisible by all numbers. He observes that 60 is divisible by 1, 2, 3, 4, 5, and 6. He

examines a few more cases, as 10, 20, and 30, taken at random as he says. Since 60 is divisible also by these, he considers the experimental evidence sufficient."

"Yes, but look at the engineers," said the physicist. "An engineer suspected that all odd numbers are prime numbers. At any rate, 1 can be considered as a prime number, he argued. Then there come 3, 5, and 7, all indubitably primes. Then there comes 9; an awkward case, it does not seem to be a prime number. Yet 11 and 13 are certainly primes. 'Coming back to 9,' he said, 'I conclude that 9 must be an experimental error.' "

HINT A.2. Cases a–c are special cases of case d.

d. Define

$$R_k(n) = 1 \cdot 2 \cdot \ldots \cdot k + 2 \cdot 3 \cdot \ldots \cdot (k+1) + \ldots + n(n+1)\ldots(n+k-1)$$

Then show that:

$$R_k(n) = \frac{1}{k+1} n(n+1)\ldots(n+k)$$

HINT A.3. Gregory's triangle gives

$$S_4(n) = 0\binom{n}{0} + 1\binom{n}{1} + 15\binom{n}{2} + 50\binom{n}{3} + 60\binom{n}{4} + 24\binom{n}{4}$$

which after some effort yields

$$S_4(n) = \frac{1}{30} n(n+1)(2n+1)(3n^2+3n-1)$$

In general the formula for $S_k(n)$ can be derived if we know expressions for $S_1(n), S_2(n), \ldots, S_{k-1}(n)$, as illustrated by the following derivation for $S_5(n)$. If in the identity:

$$(m+1)^6 = m^6 + 6m^5 + 15m^4 + 20m^3 + 15m^2 + 6m + 1$$

We set $m = 1, 2, \ldots, n$, then add all equalities obtained in this way; after canceling all sixth powers except 1^6 and $(n+1)^6$, we find

$$S_5(n) = \frac{1}{6}\left[(n+1)^6 - 1 - 15S_4(n) - 20S_3(n) - 15S_2(n) - 6S_1(n) - n\right]$$

That is:

$$S_5(n) = \frac{1}{12}n^2(n+1)^2(2n^2+2n-1)$$

HINT A.4. This result should be familiar to all readers:

$$A(n) = na + \frac{n(n-1)}{2}d = \frac{n}{2}(a_1 + a_n)$$

HINT A.5. For $q = 1$, obviously $B(n) = n$. For $q \neq 1$, consider

$$B(n) = b + bq + \ldots + bq^{n-1} \qquad qB(n) = bq + bq^2 + \ldots + bq^n$$

then write

$$qB(n) - B(n) = bq^n - bqp.$$

That is:

$$B(n) = b\frac{q^n - 1}{q-1}$$

NOTE: If $|q| < 1$ and $n \to \infty$, the last identity gives us the formula for the sum of an infinite geometric series:

$$b + bq + bq^2 + \ldots = \frac{b}{1-q}$$

HINT A.6. The identity to be proved can be guessed after calculating $n = 1, 2, 3, \ldots$ until regularity is observed. It can also be derived by using the partial fraction expansion of $1/[k(k+1)]$:

$$\frac{1}{k(k+1)} = \frac{1}{k} - \frac{1}{k+1}$$

Could we use Gregory's triangle here?

HINT A.7. This identity can be derived from:

$$1 - \frac{1}{k^2} = \frac{k-1}{k}\frac{k+1}{k}$$

HINT A.8. PROOF:

Step 1: For $n = 1$:

$$(a+b)^1 = \binom{1}{0}a^1 + \binom{1}{1}b^1 \qquad \checkmark$$

Step 2: From the inductive hypothesis:

$$(a+b)^k = \sum_{i=0}^{k} \binom{k}{i} a^{k-i} b^i$$

it follows that:

$$\begin{aligned}
(a+b)^{k+1} &= (a+b)(a+b)^k \\
&= (a+b)\sum_{i=0}^{k}\binom{k}{i}a^{k-i}b^i \\
&= \sum_{i=0}^{k}\binom{k}{i}a^{k+1-i}b^i + \sum_{i=0}^{k}\binom{k}{i}a^{k-i}b^{i+1} \\
&= \binom{k}{0}a^{k+1} + \sum_{i=1}^{k}\binom{k}{i}a^{k+1-i}b^i + \sum_{i=0}^{k-1}\binom{k}{i}a^{k-i}b^{i+1} + \binom{k}{k}b^{k+1} \\
&= \binom{k}{0}a^{k+1} + \sum_{i=1}^{k}\binom{k}{i}a^{k+1-i}b^i + \sum_{i=1}^{k}\binom{k}{i-1}a^{k+1-i}b^i + \binom{k}{k}b^{k+1} \\
&= \binom{k+1}{0}a^{k+1} + \sum_{i=1}^{k}\left[\binom{k}{i} + \binom{k}{i-1}\right]a^{k+1-i}b^i + \binom{k+1}{k+1}b^{k+1} \\
&= \sum_{i=0}^{k+1}\binom{k+1}{i}a^{k+1-i}b^i
\end{aligned}$$

This completes the proof.

HINT A.9.

a. The first four Euclidean numbers are 2, 3, 7, and 43; furthermore since we defined them by analogy to Euclid's proof of the infiniteness of the set of primes, we might think they are all prime; however $e_5 = 1807 = 13 \cdot 139$.

b. By calculating this sum for $n = 1, 2, 3, 4$ we easily guess, and then prove by induction, that:

$$\sum_{i=1}^{n} \frac{1}{e_i} = \frac{e_1 e_2 \dots e_n - 1}{e_1 e_2 \dots e_n}$$

More compactly:

$$\sum_{i=1}^{n} \frac{1}{e_i} = 1 - \frac{1}{e_{n+1} - 1}$$

c. No! For example $2 \cdot 3 \cdot 5 \cdot 7 \cdot 11 \cdot 13 + 1 = 30031 = 59 \cdot 509$.

d. Cantor's sequence begins as follows: $c_0 = 2, c_1 = 3, c_2 = 7, c_3 = 127, c_4 = 2^{127} - 1$, and they are all indeed prime. However c_5 has $5 \cdot 10^{37}$ digits, and it is not known whether it is prime or not.

HINT A.10. At one point during the course of the proof, we must leave the main problem to show (using induction or the criterion for divisibility by 3) that the number $10^{n-1} + 2$ is divisible by 3.

HINT A.11. Use the recursive relation for Fibonacci numbers:

$$f_{n+1} = f_n + f_{n-1}$$

HINT A.12. Calculating A^n for $n = 2, 3, 4, 5$ should enable the reader to discover the regularity.

HINT A.13. PROOF:

Step 1: For $n = 1$ obviously:

$$(\cos\varphi + i\sin\varphi)^1 = \cos 1 \cdot \varphi + i\sin 1 \cdot \varphi. \qquad \checkmark$$

Step 2: If we assume that:

$$(\cos\varphi + i\sin\varphi)^k = \cos k\varphi + i\sin k\varphi$$

we find

$$\begin{aligned}
(\cos\varphi + i\sin\varphi)^{k+1} &= (\cos\varphi + i\sin\varphi)(\cos\varphi + i\sin\varphi)^k \\
&= (\cos\varphi + i\sin\varphi)(\cos k\varphi + i\sin k\varphi) \\
&= (\cos\varphi\cos k\varphi - \sin\varphi\sin k\varphi) \\
&\quad + i(\sin\varphi\cos k\varphi + \sin k\varphi\cos\varphi) \\
&= \cos(k+1)\varphi + i\sin(k+1)\varphi
\end{aligned}$$

This completes the proof.

NOTE: We proved de Moivre's formula for $n \in N$. The domain of its validity is easily extended to the set of integers Z and also to the set of rational numbers Q.

As a consequence of Euler's formula:

$$e^{i\varphi} = \cos\varphi + i\sin\varphi$$

it holds for any real, even complex, value of n.

HINT A.14. Using the trigonometric identities:

$$\cos(\alpha+\beta) = \cos\alpha\cos\beta - \sin\alpha\sin\beta$$

and

$$\sin\alpha\sin\beta = \frac{1}{2}[\cos(\alpha-\beta) - \cos(\alpha+\beta)]$$

we find

$$T_{n+1}(x) = T_n(x)T_1(x) - \frac{1}{2}[T_{n-1}(x) - T_{n+1}(x)]$$

That is:

$$T_{n+1}(x) = 2T_1(x)T_n(x) - T_{n-1}(x)$$

In proving the validity of the theorem for $T_{n+1}(x)$ we used the assumption that it is true for $T_n(x)$ and $T_{n-1}(x)$; therefore in the first step of the proof, we must verify the first two cases: $T_0(x) = \cos 0 = 1$, $T_1(x) = \cos(\arccos x) = x$.

HINT A.15. For the angles $\alpha \leq \pi$:

$$2\cos\frac{\alpha}{2} = \sqrt{2+2\cos\alpha}$$

because from:

$$\cos^2\theta + \sin^2\theta = 1 \qquad \cos^2\theta - \sin^2\theta = \cos 2\theta$$

we find

$$\cos^2\theta = \frac{1+\cos 2\theta}{2}$$

HINT A.16. From the first few terms, we can see the regularity, and then prove that for $n > 0$:

$$f^{(n)}(x) = (-1)^{n-1} \frac{(n-1)!}{(1+x)^n}$$

Hence:

$$\ln(1+x) = \sum_{n=1}^{\infty} \frac{(-1)^{n+1}}{n} x^n = x - \frac{x^2}{2} + \frac{x^3}{3} - \frac{x^4}{4} + \dots$$

NOTE: Set $x = 1$ to obtain

$$1 - \frac{1}{2} + \frac{1}{3} - \frac{1}{4} + \dots = \ln 2$$

HINT A.17. PROOF:

Step 1: For $n = 1$ we have $a_1 = 1 \Rightarrow a_1 \geq 1$ $\quad\surd$

Step 2: Assume that the implication is true for $n = k$. Then if:

$$a_1 a_2 \dots a_k a_{k+1} = 1$$

then:

- If all a_i, $i = 1, 2, \dots, (k+1)$ are equal then sum is $(k+1)$, and the proof is completed.

- If one of the numbers, e.g., a_k, is > 1, at least one of the remaining numbers, e.g., a_{k+1}, must be < 1; now consider the k numbers:

  $$a_1, a_2, \dots, a_{k-1}, (a_k a_{k+1})$$

  whose product is 1, so according to the inductive hypothesis:

  $$a_1 + a_2 + \dots + a_{k-1} + a_k a_{k+1} \geq k$$

  Finally:

  $$\begin{aligned} a_1 + a_2 + \dots + a_{k-1} + a_k + a_{k+1} &\geq k - a_k a_{k+1} + a_k + a_{k+1} \\ &\geq k + 1 + (a_k - 1)(1 - a_{k+1}) \\ &> k + 1 \end{aligned}$$

This completes the proof.

Now we easily see that the equality holds if and only if all a_i, $i = 1,2,\ldots,(k+1)$ are equal. Hence the inequality of the arithmetic, geometric, and harmonic means becomes an equality if and only if all the numbers involved are equal.

Indeed if we set

$$a_i = \frac{x_i}{\sqrt{x_1 \ldots x_n}} \qquad i = 1,2,\ldots,n$$

we obtain the inequality of arithmetic and geometric means.

Similarly, if we set

$$a_i = \frac{\sqrt{x_1 \ldots x_n}}{x_i} \quad i = 1,2,\ldots,n$$

we obtain the inequality of geometric and harmonic means.

HINT A.18. The given inequality cannot be proved by induction, although the more restrictive inequality:

$$\frac{1}{2} \cdot \frac{3}{4} \cdot \ldots \cdot \frac{2n-1}{2n} \leq \frac{1}{\sqrt{3n+1}}$$

is fairly easy to prove. □

Many mathematical tricks and fallacies are based on misusing Steps 1 and 2. In Refs. [28] and [39], we can find the following examples.

EXAMPLE A.10 (WHERE IS THE ERROR?). For every $n \in N_0$, let the statement A_n be given by:

$$1 + 2 + 2^2 + \ldots + 2^n \stackrel{?}{=} 2^{n+1}$$

Since the sum at the left equals $2^{n+1} - 1$ for every $n \in N_0$, A_n is obviously always wrong. However, assume it is true for $n = k$. Then it is easy to prove it is also true for $n = k+1$:

$$\begin{aligned} 1 + 2 + 2^2 + \ldots + 2^k + 2^{k+1} &\stackrel{?}{=} 2^{k+1} + 2^{k+1} \\ &\stackrel{?}{=} 2^{k+2} \end{aligned}$$

Where is the error? We omitted the first step!

Example A.11 (All horses are the same color!). All horses are the same color! We prove that by induction on the number of horses: If we have only one horse, the proposition is true because that horse is the same color as itself. Assume now the proposition is true when we have k horses, then consider $k+1$ horses. Let these horses be marked by numbers from 1 to $k+1$. If we extract the horses with numbers 1 through k, by the inductive hypothesis they are all the same color. Similarly horses numbered 2 through $k+1$ are the same color. Finally since the relation "is the same color as" is transitive, we see that the whole group of $k+1$ horses is also the same color!

Where is the error now? This example was given in Ref. [28]. The first example similar to this one was published in Ref. [45].

Appendix B

Important Mathematical Constants

$\pi = 3.14159\ 26535\ 89793\ \ldots$

By definition the number π is the ratio of the circumference and the diameter of a circle. Archimedes was the first to prove that the same proportionality constant appears in expressions for the area of a circle, ($r^2\pi$), area of a sphere ($4r^2\pi$), and the volume of a ball ($4r^3\pi/3$), as well as in similar formulas for a cylinder and a cone.

The number π is probably the most important mathematical constant because in addition geometry, it appears in number theory, analysis, probability theory, and many other branches of mathematics.

In ancient times the Chinese thought this ratio equaled 3. The *Rhind papyrus** from Ancient Egypt (around 1650–1500 B.C.) says the area of a circle equals the square of eight-ninths of its radius, which yields $\pi \approx 256/81 = 3.16049\ldots$

Archimedes estimated the value of π in several ways. For example by calculating areas of regular 96-gons, one inscribed in the circle, the other circumscribed around it, he found that:

$$3^{10}\!/\!_{71} < \pi < 3^{10}\!/\!_{70}$$

The upper bound found by Archimedes is the popular approximation $\pi \approx 22/7$. It is the best approximation by a fraction for integers less than 100.

Ptolemy of Alexandria (second cent. A.D.) used the fraction $377/120$. Later Tsu Chung-Chi in China (fifth cent. A.D.) used $355/113 = 3.1415929\ldots$ This fraction is the best among the fractions for integers less than 30000.

This approximation was the best until Arab mathematicians found better values in the fifteenth century. At the same time, European mathematicians were far behind. They took the lead only in the sixteenth century, thanks to Ludolph van Ceulen, who determined π to 35 decimal places. In his honor π is still sometimes called the Ludolphine number. Today the value of π is known to several millions of decimal places.

The Greek letter π was first used as a symbol for this number by W. Jones in 1707, but it was not widely used until Euler began using it in 1737. Why π? Probably from the Greek words $\pi\epsilon\rho\iota\mu\epsilon\tau\rho o\nu$ (perimetron = circumference) or $\pi\epsilon\rho\iota\phi\epsilon\rho\epsilon\iota\alpha$ (periferia = periphery).

---

*This papyrus was bought in 1858 by a Scottish antiquary, A. H. Rhind, in Egypt. It is sometimes called the Ahmes papyrus in honor of the scribe who wrote it.

Probably the oldest formula that includes π comes from Viette in 1592. In today's notation, Viete's formula looks like this:

$$\frac{2}{\pi} = \sqrt{\frac{1}{2}} \cdot \sqrt{\frac{1}{2} + \frac{1}{2}\sqrt{\frac{1}{2}}} \cdot \sqrt{\frac{1}{2} + \frac{1}{2}\sqrt{\frac{1}{2} + \frac{1}{2}\sqrt{\frac{1}{2}}}} \cdots$$

Wallis famous formula was found in 1655:

$$\frac{\pi}{2} = \frac{2}{1} \cdot \frac{2}{3} \cdot \frac{4}{3} \cdot \frac{4}{5} \cdot \frac{6}{5} \cdot \frac{6}{7} \cdot \frac{8}{7} \cdots$$

Gregory (1671) and Leibniz (1673) independently found the following infinite series:

$$\frac{\pi}{4} = 1 - \frac{1}{3} + \frac{1}{5} - \frac{1}{7} + \frac{1}{9} - \ldots$$

Jakob Bernoulli asked his contemporaries for their help in summing the series:

$$\sum_{n=1}^{\infty} \frac{1}{n^2} = \frac{1}{1^2} + \frac{1}{2^2} + \frac{1}{3^2} + \frac{1}{4^2} + \ldots$$

In 1736 Euler showed that:

$$\frac{1}{1^2} + \frac{1}{2^2} + \frac{1}{3^2} + \frac{1}{4^2} + \ldots = \frac{\pi^2}{6}$$

Here are some other formulas discovered by Euler:

$$\frac{1}{1^2} + \frac{1}{3^2} + \frac{1}{5^2} + \frac{1}{7^2} + \frac{1}{9^2} + \ldots = \frac{\pi^2}{8}$$

$$\frac{2^2}{2^2-1} \cdot \frac{3^2}{3^2-1} \cdot \frac{5^2}{5^2-1} \cdot \frac{7^2}{7^2-1} \cdot \frac{11^2}{11^2-1} \cdots = \frac{\pi^2}{6}$$

The former is a sum over all odd numbers, while the latter is a product over all primes.

We mention two formulas that use continued fractions. The first of these was discovered by Brouncker in 1658:

$$\frac{4}{\pi} = 1 + \cfrac{1^2}{2 + \cfrac{3^2}{2 + \cfrac{5^2}{2 + \cdots}}}$$

$$\frac{4}{\pi} = 1 + \cfrac{1^2}{3 + \cfrac{2^2}{5 + \cfrac{3^2}{7 + \cdots}}}$$

Neither of these continued fractions is a *proper* continued fraction. They are both *generalized* continued fractions. The difference is that proper continued fractions must have all numerators equal to 1. This makes the proper continued fractions a unique representation of a number, while generalized continued fractions are not unique.

In 1733 Buffon showed that if parallel lines are drawn at a distance d between neighboring lines and a needle of length d is thrown on the paper, the probability of the needle intersecting some of the lines equals $2/\pi$.

The probability that two randomly picked integers a and b are relatively prime is approximately $6/\pi^2$. The probability that a randomly picked integer is not divisible by a square is approximately $6/\pi^2$.

The irrationality of π was proved by Lambert in 1761 and its transcendence by Lindemann in 1882. Lindemann's proof also answered the ancient question of whether a square with area equal to the area of a given circle can be constructed by ruler and compass.* The answer was of course no.

$e = 2.71828\ 18284\ 59045\ \ldots$

The number e is the base of the natural logarithms, and it is defined as:

$$\lim_{n \to \infty} \left(1 + \frac{1}{n}\right)^n$$

Directly from this definition we see that if we borrow D dollars with the annual interest of $p\%$, then assuming continuous interest compounding, after a year our debt grows to $D \cdot e^p$.

*This is one of three *classical problems*, the problem of *squaring a circle*. The other two are *doubling a cube* and the *angle trisection*. Attempts to solve these problems led to many important discoveries not only in geometry but also in algebra and number theory. (See also Example 4.53.)

Because of its many useful properties, e is often found in analysis and its applications. Its most important property, from the point of view of analysis, is $de^x/dx = e^x$.

Newton found the following expansion:

$$e = 1 + \frac{1}{1!} + \frac{1}{2!} + \frac{1}{3!} + \frac{1}{4!} + \ldots$$

This formula is a special case of the Maclaurin series expansion (actually Taylor expansion):

$$e^x = 1 + \frac{x}{1!} + \frac{x^2}{2!} + \frac{x^3}{3!} + \frac{x^4}{4!} + \ldots$$

From this expansion it is obvious that:

$$\frac{1}{e} = 1 - \frac{1}{1!} + \frac{1}{2!} - \frac{1}{3!} + \frac{1}{4!} - \ldots$$

In complex analysis we find that for any $z \in C$

$$e^z = 1 + \frac{z}{1!} + \frac{z^2}{2!} + \frac{z^3}{3!} + \frac{z^4}{4!} + \ldots$$

If we write similar series expansions for $\sin z$ and $\cos z$ and if we set $z = i\varphi$, we obtain Euler's formula* from 1743:

$$e^{i\varphi} = \cos\varphi + i\sin\varphi$$

In particular when $\varphi = \pi$, we obtain a connection between the five most important numbers in mathematics:

$$e^{i\pi} + 1 = 0$$

The notation e was first used by Euler in 1731. He was also the first to use i for the imaginary unit; before 1777 he used i for infinitely large numbers but later changed this. Using i for the imaginary unit was particularly popularized by Gauss, who used it in 1801 in his famous book *Disquisitiones Arithmeticae*.

*The equivalent formula $i\varphi = \ln(\cos\varphi + i\sin\varphi)$ was used by Cotes in 1714.

In 1737 Euler showed the irrationality of e. Its transcendence was proved by Hermite in 1873.

Let us denote a *proper* continued fraction

$$a_0 + \cfrac{1}{a_1 + \frac{1}{a_2 + \cdots}}$$

by $[a_0; a_1, a_2, \ldots]$, then for e we can write

$$e = [2; 1, 2, 1, 1, 4, 1, 1, 6, 1, 1, 8, 1, 1, \ldots]$$

$\gamma = 0.57721\ 56649\ 01533\ \ldots$

Euler's name is associated with this constant, too, the so-called *Euler's constant.* By definition,*

$$\gamma = \lim_{n \to \infty} \left(1 + \frac{1}{2} + \frac{1}{3} + \ldots + \frac{1}{n} - \ln n\right)$$

It is still not known whether this number, introduced by Euler in 1781, is a rational or an irrational number.

If $2, 3, 5, 7, \ldots, p$ are all primes $\leq N$, Mertens' asymptotic formula holds

$$\left(1 + \frac{1}{2}\right)\left(1 + \frac{1}{3}\right)\left(1 + \frac{1}{5}\right)\left(1 + \frac{1}{7}\right) \ldots \left(1 + \frac{1}{p}\right) \sim \frac{6e^{\gamma}}{\pi^2} \ln N$$

$\phi = 1.61803\ 39887\ 49894\ \ldots$

The golden section is certainly the most unusual number, because it appears not only in mathematics but also in astronomy, biology, psychology, arts, and architecture. By definition, the golden section is the ratio of the greater and the smaller parts in the division of the line segment $[AB]$ by the point $M \in [AB]$ (see Fig. B.1), picked so that:

$$[AB] : [AM] = [AM] : [MB]$$

*The basis for this definition is the fact that the harmonic series diverges, harmonic numbers $H_n = 1 + \frac{1}{2} + \frac{1}{3} + \ldots + \frac{1}{n}$ asymptotically behave as $H_n \sim \ln n$, and for $n \to \infty$ the limit $\lim(H_n - \ln n)$ exists.

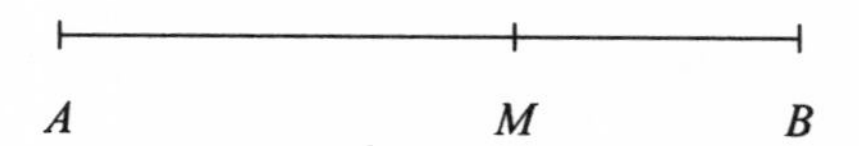

FIGURE B.1. Definition of the golden section. If $[AB]:[AM]=[AM]:[MB]$, then $\phi=[AM]:[MB]$.

Then

$$\phi=[AM]:[MB]$$

From this definition we find that:

$$\phi^2-\phi-1=0 \qquad \phi>1 \tag{B.1}$$

That is:

$$\phi=\frac{1+\sqrt{5}}{2}$$

Obviously ϕ is an algebraic irrational number, since it is a root of an algebraic equation with rational coefficients.

NOTE: The other root of the quadratic equation (B.1) is

$$\hat{\phi}=\frac{1-\sqrt{5}}{2}=-\frac{1}{\phi}$$

Euclid used the golden section in his *Elements* to construct a regular pentagon and the Platonic solids (regular polyhedra), which have regular pentagons in them, dodecahedron and icosahedron. The relation of the golden section to the regular pentagon and the regular decagon is condensed in the following trigonometric identity:

$$\phi=2\cos\frac{\pi}{5}$$

In Ancient Greece it was simply called the *section*. During the Renaissance the artists, mystics, and scientists called it the *divine proportion*, until Leonardo da Vinci gave it its modern name, *sectio aurea*, the golden section.

There is an interesting relation between the golden section and the sequence of Fibonacci numbers. In 1718 de Moivre found an explicit formula for

Fibonacci numbers, known today as Binet's formula:

$$f_n = \frac{\sqrt{5}}{5}\left[\left(\frac{1+\sqrt{5}}{2}\right)^n - \left(\frac{1-\sqrt{5}}{2}\right)^n\right] = \frac{\sqrt{5}}{5}\left(\phi^n - \hat{\phi}^n\right)$$

Thanks to this formula we see that ratios of consecutive Fibonacci numbers tend to the golden section:

$$\lim_{n \to \infty} \frac{f_{n+1}}{f_n} = \phi$$

Until recently the golden section was denoted by τ, the first letter of its Ancient Greek name $\tau o \mu \eta$ (tome = section), or by g; today the most common symbol is ϕ, in honor of the Ancient Greek sculptor Phidias, who thought golden proportions are the most pleasing to the human eye. This notation was introduced by Mark Barr and widely popularized by Donald Knuth.

From the quadratic equation $x^2 - x - 1 = 0$, we easily find that

$$\phi = \sqrt{1 + \sqrt{1 + \sqrt{1 + \ldots}}}$$

The golden section has a special place in the theory of continued fractions because:

$$\phi = [1; 1, 1, 1, 1, 1, 1, \ldots]$$

It is interesting that the approximation of ϕ by a continued fraction of order n is equal to the ratio of consecutive Fibonacci numbers f_{n+2} and f_{n+1}; for example:

$$[1; 1, 1, 1, 1, 1] = \frac{13}{8} = \frac{f_7}{f_6}$$

Appendix C

Great Mathematicians

FIGURE C.1. Archimedes, Newton, Euler, and Gauss were probably the greatest mathematicians of all time.

Archimedes of Syracuse was the most prominent mathematician and inventor of Ancient Greece (see Fig. C.1). He was born between 290 and 280 B.C. in Syracuse, on the island of Sicily. There at the court of king Hieron II, he spent most of his life. He died during the Roman siege of Syracuse at the end of 212 or at the beginning of 211 B.C. Several physical laws and mathematical theorems bear his name. There are many legends and anecdotes about his life; the most popular is about his discovery of how to verify whether the king's crown was made of pure gold. The physical law explaining Archimedes' idea is still known as the law of Archimedes. According to legend, when Archimedes made the discovery, he ran out from his bath exclaiming *"Heureka!"* His works had a great impact on the development of mathematics, especially in the sixteenth and the seventeenth century, when they became available (through Arab translations) in Europe. His influence would probably be even greater if some of his works, especially *Method Concerning Mechanical Theorems*, had not been lost until the beginning of the twentieth century.

Sir Isaac Newton was an English physicist and mathematician (see Fig. C.1). He was born on January 4, 1643 (December 25, 1642 old style) in Woolsthorpe, and he died on March 31, 1727 in London. His discoveries in optics, mechanics, theory of gravitation, as well as the discovery of the infinitesimal calculus caused a revolution in natural sciences and mathematics. His books *Opticks*, and in particular *Philosophiae Naturalis Principia Mathematica*, rank among the most important and most influential works of modern science. Newton was a man of unpleasant character, with no tolerance for opinions different than his own. Famous are his discussions with Leibniz over priority in the discovery of calculus. Today historians agree that Newton was the first to discover it, but Leibniz discovered it independently, and he was the first to publish it. Therefore their mutual accusations of plagiarism were completely unfounded.

Leonhard Euler was probably the most productive and versatile mathematical mind of all time, considering not only the volume of his works, but also the number of disciplines he improved or influenced (see Fig. C.1). An incredibly large number of theorems and formulas still bear his name today. A large portion of the mathematical notation we use today was introduced by him. He was born on April 15, 1707 in Basel, Switzerland, and he died on September 18, 1783 in St. Petersburg, Russia, where he spent most of his life. Euler's father, a Calvinist pastor, was a student of Jakob Bernoulli, while Euler was a student of Jakob's brother, Johann. Euler obtained a position at the St. Petersburg Academy of Sciences, after he was recommended by Johann Bernoulli and his sons, Daniel and Nicolaus (II), who also worked there. For many years Euler also lived in Berlin. He had 13 children, only five of whom survived early childhood. His enormous productivity was not disturbed even by the fact that he had been blind in one eye since 1735, and since 1766 in both eyes!

Carl Friedrich Gauss, a great German mathematician, was born on April 30, 1777 in Brunswick, and he died on February 23, 1855 in Göttingen (see Fig. C.1). His exceptional talent for mathematics and languages attracted the attention of the Duke of Brunswick, who provided generous support for Gauss's education, which was crowned by a doctorate in 1799. Gauss chose a mathematical career over linguistics when he was 19, after solving the ancient problem of constructing a regular heptagon by proving that its construction is impossible if only ruler and the compass are used. In his doctoral thesis, Gauss gave the first proof of the *Fundamental Theorem of Algebra*, i.e., every polynomial has at least one complex root. This proof was earlier attempted by Euler, D'Alembert, Laplace, and Lagrange. Besides the contributions we already mentioned, Gauss greatly advanced number theory, and he had equally great influence on applied mathematics, electromagnetism, astronomy, and geodesy. Many mathematical techniques still used today in science and engineering, such as the method of elimination in solving systems of linear equations, the least-squares method in statistical data analysis, even the Fast Fourier Transform (FFT) algorithm, were first used by Gauss. He had six children, and he lived primarily in Göttingen.

Important Names from the History of Mathematics

THALES OF MILETUS (SIXTH CENTURY B.C.)
PYTHAGORAS OF SAMOS (SIXTH CENTURY B.C.)
EUDOXUS OF CNIDUS (FOURTH CENTURY B.C.)
PLATO OF ATHENS (FOURTH CENTURY B.C.)
EUKLID OF ALEXANDRIA (FOURTH CENTURY B.C.)
ARCHIMEDES OF SYRACUSE (THIRD CENTURY B.C.)
ERATOSTHENES OF CYRENE (THIRD CENTURY B.C.)
APOLLONIUS OF PERGA (THIRD CENTURY B.C.)
HERO OF ALEXANDRIA (FIRST CENTURY A.D.)
SUN-TSU (FIRST CENTURY A.D.)
PTOLEMY OF ALEXANDRIA (SECOND CENTURY A.D.)
DIOPHANTUS OF ALEXANDRIA (THIRD CENTURY A.D.)
PAPPUS OF ALEXANDRIA (FOURTH CENTURY A.D.)
BRAHMAGUPTA (598–665)
MUHAMED IBN MUSA AL-HOVARIZMI (780–850)
LEONARDO FIBONACCI (LEONARDO PISANO) (1170–1240)
LEONARDO DA VINCI (1452–1519)
NICCOLO TARTAGLIA (1500–1557)
GEROLAMO CARDANO (1501–1576)
LUDOVICO FERRARI (1522–1565)
FRANCOIS VIETE (1540–1603)
JOHN NAPIER (1550–1617)
HENRY BRIGGS (1561–1630)

Marin Marsenne (1588–1648)
Rene Descartes (1596–1650)
Pierre de Fermat (1601–1665)
John Wallis (1616–1703)
Blaise Pascal (1623–1662)
Christiaan Huygens (1629–1695)
James Gregory (1638–1675)
Isaac Newton (1643–1727)
Gottfried Wilhelm Leibniz (1646–1716)
Giovanni Ceva (1648–1734)
Jakob Bernoulli (1655–1705)
Abraham de Moivre (1667–1754)
Johann Bernoulli (1667–1748)
Brook Taylor (1685–1731)
Robert Simson (1687–1768)
James Stirling (1692–1770)
Colin Maclaurin (1698–1746)
Daniel Bernoulli (1700–1782)
Leonhard Euler (1707–1783)
Matthew Stewart (1717–1785)
Jean Le Rond d'Alambert (1717–1783)
Alexandre Vandermonde (1735–1796)
Joseph-Louis Lagrange (1736–1813)
John Wilson (1741–1793)
Pierre-Simon Laplace (1749–1827)
Adrien-Marie Legendre (1752–1833)
Jean-Baptiste-Joseph Fourier (1768–1830)
Carl Friedrich Gauss (1777–1855)
Simon-Denis Poisson (1781–1840)
Jacques Binet (1786–1856)
Augustin-Louis Cauchy (1789–1857)
August Ferdinand Möbius (1790–1868)
Nikolay Ivanovich Lobachevsky (1792–1856)
Jakob Steiner (1796–1863)
Karl Feuerbach (1800–1834)
Niels Abel (1802–1829)
Janos Bolyai (1802–1860)
Karl Gustav Jakob Jacobi (1804–1851)

PETER GUSTAV LEJEUNE DIRICHLET (1805–1859)
WILLIAM ROWAN HAMILTON (1805–1865)
AUGUSTUS DE MORGAN (1806–1871)
JOSEPH LIOUVILLE (1809–1882)
EVARISTE GALOIS (1811–1832)
JAMES SYLVESTER (1814–1897)
GEORGE BOOLE (1815–1864)
PAFNUTY LVOVICH CHEBYSHEV (1821–1894)
ARTHUR CAYLEY (1821–1895)
BERNHARD RIEMANN (1826–1866)
GEORG CANTOR (1845–1918)
FELIX KLEIN (1849–1925)
HENRI POINCARE (1854–1912)
ANDREY ANDREYEVICH MARKOV (1856–1922)
ALEKSANDR MIHAILOVICH LYAPUNOV (1857–1918)
GIUSEPPE PEANO (1858–1932)
DAVID HILBERT (1862–1943)
HERMANN MINKOWSKI (1864–1909)
BERTRAND RUSSEL (1872–1970)
GODFREY HAROLD HARDY (1877–1947)
SRINIVASA RAMANUJAN (1887–1920)
GEORGE POLYA (1887–1985)
NORBERT WIENER (1894–1964)
ANDREY NIKOLAYEVICH KOLMOGOROV (1903–1987)
JOHN VON NEUMANN (1903–1957)
KURT GÖDEL (1906–1978)
ALAN TURING (1912–1954)

Biographies of these and many other mathematicians can be found on the internet, for example, at: `www-groups.dcs.st-andrews.ac.uk/~history`

Appendix D

Greek Alphabet

α	A	alpha
β	B	beta
γ	Γ	gamma
δ	Δ	delta
ε	E	epsilon
ζ	Z	zeta
η	H	eta
θ	T	theta
ι	I	iota
κ	K	kappa
λ	Λ	lambda
μ	M	mu
ν	N	nu
ξ	Ξ	xi
o	O	omicron
π	Π	pi
ρ	P	rho
σ	Σ	sigma
τ	T	tau
υ	Y	upsilon
ϕ, φ	Φ	phi
χ	X	chi
ψ	Ψ	psi
ω	Ω	omega

References

1. Aaboe, A., *Episodes from the Early History of Mathematics* (Random House, New York, 1964).
2. Balk, M. B., and V. G. Boltijanskij, *Geometriia mass* (Bibliotechka Kvant, Nauka, Moskva, 1987).
3. Bashmakov, M. I., B. M. Bekker, and V. M. Goljhovoj, *Zadachi po matematike* (Bibliotechka Kvant, Nauka, Moskva, 1982).
4. Bell, E. T., *Men of Mathematics* (Touchstone, New York, 1937).
5. Bold, B., *Famous Problems of Geometry and How to Solve Them* (Dover, New York, 1969).
6. Born, M., *Atomic Physics*, 8th ed. (Dover, New York, 1989).
7. Brualdi, R. A., *Introductory Combinatorics*, 2nd ed. (Prentice Hall, Englewood Cliffs, 1992).
8. Bryant, V., *Aspects of Combinatorics* (Cambridge University Press, 1993).
9. Bullen, P. S., D. S. Mitrinović, and P. M. Vasić, *Means and their Inequalities* (D.Reidel, Dodrecht, 1988).
10. Carroll, L., *Lewis Carroll Picture Book* (Tower Books, Detroit, 1971).
11. Cohen, D. I. A., *Basic Techniques of Combinatorial Theory* (Wiley, New York, 1978).
12. *Collier's Encyclopedia* (Collier, New York, 1992).
13. Conway, J. H., and R. K. Guy, *Book of Numbers* (Copernicus, New York, 1995).
14. Cormen, T. H., C. E. Leiserson, and R. L. Rivest, *Introduction to Algorithms* (MIT Press, Cambridge, MA, 1990).
15. Courant, R., and H. Robbins, *What is Mathematics?* 2nd ed. (Revised by I. Stewart) (Oxford University Press, 1996).
16. Coxeter, H. S. M., and S. L. Greitzer, *Geometry Revisited* (Random House, New York, 1967).
17. Coxeter, H. S. M., *Introduction to Geometry* (Wiley, New York, 1969).
18. Cvetković, D., *Teorija grafova i njene primene* (Naučna knjiga, Belgrade, 1981).
19. Cvetković, D., and S. Simić, *Diskretna matematika* (Naučna knjiga, Belgrade, 1990).
20. Dörrie, H., *100 Great Problems of Elementary Mathematics, Their History, and Solutions* (Dover, New York, 1965).

21. Davis, D. M., *Nature and Power of Mathematics* (Princeton University Press, Princeton, NJ, 1993).
22. Devlin, K., *Mathematics: New Golden Age* (Penguin, London, 1988).
23. Dijksterhuis, E. J., *Archimedes* (Princeton University Press, Princeton, NJ, 1987).
24. Doroslovački, R., *Algebra* (Stylos, Novi Sad, 1995).
25. Euclid, *Thirteen Books of Euclid's Elements* (T. L. Heath, ed., Dover, New York, 1956).
26. Eves, H., *A Survey of Geometry, Vol. 1* (Allyn and Bacon, Boston, 1965).
27. Gardner, M., *My Best Mathematical and Logic Puzzles* (Dover, New York, 1994).
28. Graham, R. L., D. E. Knuth, and O. Patashnik, *Concrete Mathematics: A Foundation for Computer Science* (Addison-Wesley, Reading, MA, 1989).
29. Heath, T. L., *A Manual of Greek Mathematics* (Dover, New York, 1963).
30. Hogben, L., *Mathematics for the Million* (Pan Books, London, 1978).
31. Hoggatt, V. E., Jr., *Fibonacci and Lucas Numbers* (Houghton Mifflin, Boston, 1969).
32. Huntley, H. E., *Divine Proportion* (Dover, New York, 1970).
33. Kac, M., and S. M. Ulam, *Mathematics and Logic* (Dover, New York, 1992).
34. Khintchine, A. Ya., *Continued Fractions* (Nordhoff, Groningen, 1964).
35. Klein, F., *Famous Problems of Elementary Geometry* (Hafner, New York, 1950).
36. Knuth, D. E., *Art of Computer Programming*, 2nd ed., *Vol. 1, Fundamental Algorithms* (Addison-Wesley, Reading, MA, 1973).
37. M. Petković, *Zanimljivi matematički problemi* (Naučna knjiga, Belgrade, 1988).
38. Mićić, V., and Z. Kadelburg, *Uvod u teoriju brojeva* (DM SRS, Belgrade, 1983).
39. Mitrinović, D. S., *Matematička indukcija, binomna formula i kombinatorika* (Građevinska knjiga, Belgrade, 1990).
40. Mladenović, P., *Kombinatorika* (Drugo izdanje, DMS, Belgrade, 1992).
41. Neugebauer, O., *Exact Sciences in Antiquity* (Princeton University Press, Princeton, NJ, 1952).
42. *New Encyclopaedia Britannica*, 15th ed. (Encyclopaedia Britannica, Chicago, 1991).
43. Ore, O., *Number Theory and Its History* (McGraw-Hill, New York, 1948).
44. *Oxford English Dictionary*, 2nd ed. (Clarendon Press, Oxford, 1989).
45. Pòlya, G., *Induction and Analogy in Mathematics, Vol. 1 of Mathematics and Plausible Reasoning* (Princeton University Press, Princeton, NJ, 1954).
46. Rademacher, H., and O. Toeplitz, *Enjoyment of Math* (Princeton University Press, Princeton, NJ, 1957).
47. Ribenboim, P., *Book of Prime Number Records* (Springer-Verlag, Berlin, 1989).
48. Riordan, J., *Generating Functions*, in *Applied Combinatorial Mathematics* (E. F. Beckenbach, ed., Wiley, New York, 1964).
49. Rouse Ball, W. W., and H. S. M. Coxeter, *Mathematical Recreations and Essays* (Macmillan, New York, 1944).
50. Scharlau, W., and H. Opolka, *From Fermat to Minkowski* (Springer-Verlag, Berlin, 1985).
51. Schroeder, M. R., *Number Theory in Science and Communication* (Springer-Verlag, Berlin, 1986).

52. Sierpinski, W., *Elementary Theory of Numbers* (North-Holland, Amsterdam, 1988).

53. Slomson, A., *An Introduction to Combinatorics* (Chapman and Hall, London, 1991).

54. Smith, D. E., *History of Mathematics, Vol. 1* (Ginn and Company, Boston, 1923).

55. Stewart, I., *Galois Theory*, 2nd ed. (Chapman & Hall, London, 1989).

56. Struik, D., *A Concise History of Mathematics* (Dover, New York, 1967).

57. Tošić, R., and V. Vukoslavčević, *Elementi teorije brojeva* (Alef, Novi Sad, 1995).

58. Vajda, S., *Fibonacci and Lucas Numbers, and the Golden Section* (Ellis Horwood, Chichester, UK, 1989).

59. van der Waerden, B. L., *Science Awakening* (Noordhoff, Groningen, 1954).

60. Vaught, R. L., *Set Theory: An Introduction*, 2nd ed. (Birkhäuser, Boston, 1995).

61. Veljan, D., *Kombinatorika s teorijom grafova* (Školska knjiga, Zagreb, 1989).

62. Wells, D., *Penguin Dictionary of Curious and Interesting Numbers* (Penguin Books, London, 1987).

63. Wells, D., *Penguin Book of Curious and Interesting Puzzles* (Penguin Books, London, 1992).

64. Wiles, A., *Annals of Mathematics*, **141**, No. 3, pp. 443–551 (May 1995).

65. Yaglom, A. M., and I. M. Yaglom, *Challenging Mathematical Problems with Elementary Solutions, Vol. 1 and 2* (Dover, New York, 1987).

66. Yakovlev, G. N., ed., *High-School Mathematics* (Translated from the Russian) (Mir Publishers, Moscow, 1984).

Index